Think like a SCIENTIST

Think like a SCIENTIST

Explore the extraordinary natural laws of the universe

Anne Rooney

This edition published in 2020 by Arcturus Publishing Limited
26/27 Bickels Yard, 151–153 Bermondsey Street,
London SE1 3HA

AD007328UK

Printed in the UK

CONTENTS

Why does science matter?

The word 'science' comes from the Latin *scientia*, from *scire*, 'to know'. Science in its true sense is not restricted to any set of subjects, but is all knowledge.

Knowing about science means knowing about the world around you – and even the world within you – and how it works. It is far more wide-ranging than the science we associate with the school curriculum.

A long history

Science is an endeavour that began at least as long ago as the Ancient Greeks, and in some forms in Mesopotamia 4,000 years ago. But it hasn't all been plain sailing. In some parts of the world there have been long periods during which science was shunned or even forbidden. During these times people adopted different organizing principles, including the invocation of spiritual and mystical explanations for phenomena. Often the notion that only certain types of knowledge should be sought has stifled scientific inquiry.

Science is characterized by a particular way of organizing knowledge. This way developed in the 17th and 18th centuries,

Ancient traditions: in this wall panel from 865–860BC, King Ashurnasirpal appears twice on either side of a Sacred Tree, possibly symbolizing life.

when people became interested in discovering the natural laws that govern the behaviour of the physical universe.

The scientific method

The scientific method that lies at the heart of modern science developed during the Enlightenment, a period of renewed interest and confidence in inquiry into the natural and physical world. The method is rooted in empiricism – that is, what can be objectively observed and tested, rather than approached only through reason.

An idea – a theory or hypothesis – emerges from observations of the world and from thoughts about why it might be as it is. Perhaps someone observed that plants grow better at the sunny end of a field than the shady end, and so proposed that sunlight helps plants to grow. The hypothesis could then be tested by experiment or structured observations, and the results examined dispassionately and objectively to determine whether or not they support the hypothesis. Most of us met this method in school science lessons with simple experiments. The hypothesis might have been that sugar dissolves more quickly in hot water than in cold water, or that a truck will run faster down a steep slope than a shallow one. But in a school science lesson, the teacher and the textbook (and often the pupil) already have the answer. In real-world

A HYPOTHESIS MUST BE FALSIFIABLE

It may seem a strange way of looking at things, but whether or not a hypothesis is valid depends on whether it can be proved to be wrong. A hypothesis such as all dogs are more than 5cm tall can be proved wrong by finding a smaller dog, but it can't be proved right – we would need to examine every dog that has ever lived or will ever live.

science, the answer is not generally known. Some hypotheses are very tentative and proven to be wrong. Others look fairly obvious, but must still be tested before they can be treated as an accurate representation of fact.

It's perfectly possible to come to the wrong conclusion about the state of things just by looking at them. Before the invention of the microscope, around 1600, people reasonably assumed that the smallest living things were tiny bugs, such as fleas. We now know that most of the living things in the world are much smaller than this and can only be seen with the aid of a microscope, hence the term 'microscopic'. The way we look at the world and the deductions we make about it have been changed by the invention of the microscope.

Why know about science?

Curiously, it became rather trendy in the late 20th century to claim to know nothing about science or mathematics. There was a popular belief that scientific knowledge was somehow at odds with being a cultured individual, when in fact it is central to it. In 1959, the English scientist and novelist C.P. Snow delivered a famous lecture in which he spoke of the distance and even animosity that existed between the 'two cultures' of science and the arts, or humanities. This deep division in intellectual life was, he felt, holding up human progress. The divide remained, and even increased, in the following decades. It might now be narrowing, but it's far from closed.

More people, though, now recognize that knowing about science is not a mark of philistinism, but the very opposite. An informed appreciation of the world around us, the laws it follows and how we can discover those laws, puts us in the best position to make the most of our individual lives and the resources the planet affords us as a species. The loss of wild places was mourned poetically by 19th-century writers such as Wordsworth or Thoreau and demonstrated a need to find out how to renew and protect the environment. But this remorse for what humans have done to the world, invoked through the arts, can be harnessed in helpful action through the application of science and understanding.

Be prepared

Understanding something of science enables us to make informed decisions and protects us from the deceptions practised by large corporations, the media and national governments against an uninformed public. A little knowledge can protect you from scare-mongering and scams as well as raise your engagement in and wonder at the world around you.

This book can't hope to cover all the aspects of science that will equip you to understand what lies behind every news story or topical issue. But it can lead you to think more carefully about the stories you encounter and about the natural world. It can encourage an interrogative, curious or 'scientific' approach and a respect for the disciplined organization of knowledge that underpins the modern world.

> ***'So the great edifice of modern physics goes up, and the majority of the cleverest people in the western world have about as much insight into it as their Neolithic ancestors would have had.'***
>
> C.P. Snow

CHAPTER 1

Are humans the pinnacle of evolution?

We like to think we're at the top of the evolutionary tree – but are we? And is there even a tree to climb to the top of?

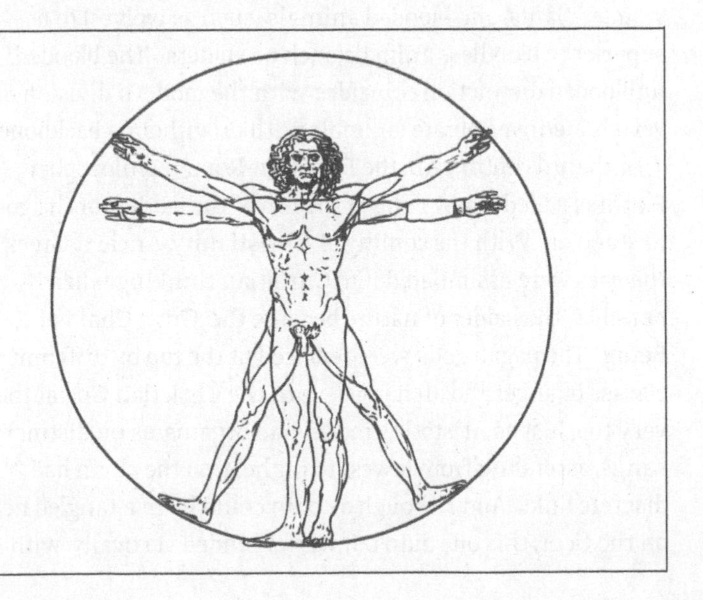

Ladder, chain or tree?

More than two thousand years ago, the Greek philosopher Aristotle wrote about the *scala naturae*, or 'ladder of Nature.' He ranked organisms (living things) in a hierarchical order, from the lowliest – simple plants – to the most advanced – human beings. This wasn't based just on feeling superior to mushrooms or flatfish. He proposed that organisms have different types of souls according to their nature and needs. The soul, he claimed, gives the physical matter of the body its capabilities.

According to Aristotle, a plant has a soul capable only of growth and sustaining life, but an animal has a soul capable of growth, sustaining life, and moving around. A human is better still, as it can do all those and is also capable of reason. For the Ancient Greeks, the rational soul placed humans at the top of the ladder.

Aristotle distinguished within the broad categories of plant/animal/human, too. He considered trees to be superior to smaller plants, and blooded animals (such as wolves) to be superior to bloodless animals(such as spiders). The blooded/unblooded distinction coincides with the modern division of vertebrate/invertebrate (animals with or without a backbone).

In the 3rd century AD, the Egyptian-Roman philosopher Plotinus added a new rung at the top of the ladder for the gods to stand on. With the coming of Christianity, Ancient Greek theories were assimilated into Christian thinking where possible. The ladder of nature became the 'Great Chain of Being'. The pagan gods were replaced at the top by different classes of angel and archangel, with the Christian God at the very top. Just as Aristotle's model had organisms on distinct rungs, ascending from lowest to highest, so the chain had discrete links. And although a chain could lie in a tangled heap on the floor, this one didn't. It was extended vertically, with angels at the top and the lowest organisms – algae, perhaps – at the bottom.

Far more organisms were known in the Middle Ages than had been familiar to Aristotle, and more were constantly being discovered over the following centuries as European adventurers, explorers and conquerors travelled further afield. The Americas, Asia, the Pacific Islands and Australasia all yielded new beings that had to be fitted into the chain, and they were. The prevalent belief was that Creation was full – God had created a perfect world, with an organism to occupy every niche, leaving no gap unfilled, even if people had not yet found all the organisms.

In a chain, everything is linked. Instead of taking a step up from plants to animals, as on a ladder, the chain model proposed intermediate links. These could be represented by organisms that were thought to share aspects of both – so shellfish or sponges that don't move are on the border between plants and animals. But some odd hybrids were also described, such as barnacle geese, which were thought to grow on trees.

No change

Both models of a ladder and of a chain describe a static order. The Abrahamic religions make the fixity of nature explicit. The account of Creation given in Genesis is of God creating the plants, then the animals and, finally, humans. The other organisms were created purely to serve humankind, so humankind is clearly their superior. Just as important as humans' superiority is the idea that all organisms have existed from the start. Creation was both perfect and complete: the world did not change and had not ever changed. How could it change, if God had created a perfect world?

The discovery of fossils challenged that notion. The fossils

BARNACLE GEESE – HATCHED FROM BARNACLES, GROWN ON TREES

How could a society with no notion of migratory birds account for the fact that they never saw barnacle goslings or saw the parents sitting on a nest? They concluded that the geese hatched from barnacles. Barnacles are often found on timber in the sea. Clearly the wood had fallen into the sea with the goose-pods growing. Perhaps they were produced from the sap in the wood. The barnacles hung down from timbers in their hundreds and – when no one was looking – grew to maturity and the geese flew away or dropped into the sea to swim off. Now we know that barnacle goslings hatch from eggs, like any other bird.

of sea creatures were found far inland, even on hills and mountains. And then, starting in earnest at the beginning of the 19th century, people began to uncover fossils of animals very unlike those that were alive at the time. First the fossil remains of plesiosaurs and ichthyosaurs were discovered, then Iguanodon and Hadrosaurus followed. The conclusion that large, unfamiliar animals had once walked (and swum) the Earth became irresistible for many scientists, though others clung to the Creation narrative and tried to explain away the discoveries. In the second half of the 19th century and the first years of the 20th, the great dinosaurs were discovered – Apatosaurus, Tyrannosaurus, Stegosaurus, Triceratops – and the human view of the past changed forever.

Evolution evolving

The theory of evolution did not spring fully formed from nowhere (or from Darwin's brain) in the mid-19th century. Even before Aristotle, some of the Greeks had proto-evolutionary ideas.

Anaximander (c.611–547BC) proposed that the first animals were formed from bubbling mud. They had lived in the water at first, but as the land and water separated over time, some of them adapted to living on land. He believed that even humans had developed from earlier, fish-like animals. After a promising start, though, western thought became bogged down in the no-change theory of Creation. Evolutionary ideas did not re-emerge in the West for around 2,000 years.

From the 18th century, evidence that organisms do change was piling up. Greater interest in taxonomy, especially after the work of the Swedish naturalist Carl Linnaeus (1707–78), showed there were clear similarities between some species widely separated by geography. Camels and llamas are similar, as are jaguars and leopards, yet there is an ocean between their territories. Scientists attempted to explain these puzzles first within the framework of traditional Christian thinking. Perhaps organisms had started off perfect but degenerated over time. Or, if they all started out in the far north and moved southwards, that would explain how animals in the New World and Old World could resemble one another – both the llama and the camel could have degenerated as they travelled through time and terrain. Genesis made allowance for degeneration, in a way, since the Fall of Man had sullied Creation. Humankind could still be the top of the heap of non-angelic beings, even if the lower rungs or links shifted a little.

The French naturalist Jean-Baptiste Lamarck (1744–1829) suggested that rather than degenerating, organisms underwent improvement, or at least adaptation. He believed that change came about as animals strove to survive. So, for example, an animal such as a giraffe that was constantly trying to reach the juiciest leaves high up in a tree would stretch its neck in the process. The results of this stretching were passed on to the next generation, so over time the animal's descendants would have longer and longer necks. In this way, striving was piled upon inheritance in a continuous sequence.

Erasmus Darwin (1731–1802), a contemporary of Lamarck, shared this view. He suggested that all life had evolved from a common ancestor over a very long period of time. The history of life could be seen as a 'single filament' joining past and present, He also proposed the idea of sexual selection. Many animals compete for mates, so: 'The final course of this contest among males seems to be that the strongest and most active animal should propagate the species which should thus be improved.'

Evolution to the fore

In 1831, Erasmus's grandson, Charles Darwin (1809–82), was just 22 years old when he set off on a journey around the world on the surveying ship HMS *Beagle*. He held the position of official naturalist for the voyage, and over the next four years and nine months he would collect samples of plants, animals

Species similarities: camels are an Old World animal living in North Africa, the Middle East and across Mongolia. Llamas are the related New World version, found in South America.

and fossils, make copious notes, observe animals and plants in their natural habitats, and wonder at both the diversity and the remarkable similarities he witnessed in the natural world. Whenever he arrived at port his latest discoveries would be boxed up and sent back to be studied and marvelled at by the scientific community back in England. By the time he returned, he had become a renowned scientist. But he did not turn with any great resolve to the question of how that diversity and alikeness had developed until 1838, and only began writing in earnest in 1842. It took him until 1859 to finish and publish what became a world-changing book, *On the Origin of Species by Means of Natural Selection.*

Darwin not only set out to assert that species change over time but also to explain why they do. He cites the method of artificial selection which farmers and pigeon-fanciers use to breed animals (or plants) that have the features they want. Breeding selectively reinforces the desired traits. Nature, Darwin said, does the same. But in nature this selection serves to make organisms better adapted to their lifestyles and environments, and not more useful or attractive to humans. An adaptation that makes an animal better able to find food, more attractive to potential mates or better able to cope with different habitats is likely to be reinforced over time. Darwin called this 'variation by natural selection'. Over time, species change through this process of variation and entirely new species develop. The idea that during the Creation animals and plants were produced purely for the use and interests of humankind was toppled. Darwin showed that organisms served their own ends. Where did that leave humanity?

What's the point?

Evolution in the Darwinian model does not have an end goal. Organisms do not develop features in order to do something, but fortuitous variations that suit an organism to do something useful are likely to be retained and reinforced over the generations. Similarly, old features which are no longer of use, like the snake's legs, are discarded.

From generation to generation there will appear many different variations between members of the same species, some of which will be disadvantageous. A lizard born with no eyes will be less successful at finding food in daylight than a sighted lizard. But if a lizard starts to inhabit a totally dark environment, such as a deep cave (as some do), the effort put into growing and maintaining eyes is wasted. Eyes may even be a liability in this situation because they are vulnerable to injury. So a lizard born without eyes might be more successful. Even so, evolution does not have a goal, it 'stumbles in the right direction' – to borrow a phrase from one of Darwin's early critics.

Who's at the top?

The early models of the ladder of nature and the great chain of being organized the natural world into a hierarchy.

Darwin sketched the structure of evolution as a tree, with many branches that divide again and again as new species split off from old ones. Even he, though, tended to put humans in a dominant position. Humans sit naturally at the top of the tree. In fact, every successful organism is at the end of a branch, and just as no twig on a tree is more important than any other twig, so no evolved organism is 'better' than any other. This is difficult if not impossible to reconcile with a view in which the natural world is considered the creation of a supernatural being with a particular interest in humankind.

Today the evolutionary relationships between organisms are depicted as cladograms. These show where significant divergences from an evolutionary path have produced a new branch. Every group of organisms is shown on the same horizontal level so there is no suggestion that one is more developed than another. To represent the entire biosphere, the cladogram is shown as a circle with the organisms all around the edge.

Where is the top?

The sense that humans are the most advanced or most evolved organism comes from the fact that we are judging what is advanced or worthwhile by human standards. We prize intelligence and think ourselves the most intelligent animals. Our definition of intelligence is based on human values and achievements, though, so it's rather a self-justifying claim: humans are most intelligent, because intelligence is the most human capacity. By a different measure of intelligence, dolphins or whales might be considered smarter than us. They don't build complex cities or sophisticated tools, we are not aware of them having music, literature or philosophy (though it's not necessarily the case that they don't), and their physiology is such that it would be hard for them to make, say, electronic circuits. (And it doesn't help that they live underwater – an environment not compatible with electricity.) But we have no idea what cetacean achievements might be or how they are valued. By cetacean standards, a species that despoils its environment and kills its peers for no good reason might rank very low on the intelligence scale.

But why use intelligence as the standard at all? If we rated organisms by efficient locomotion, longevity or ability to fly, humans wouldn't do very well. Similarly, if we judged 'evolvedness' by whether an organism is fit for survival in its environment, or the length of time a species has remained successful, humans would not rank very high. Cyanobacteria are extremely simple in physiological terms, but have been around for 3.5 billion years and so are true survivors. Modern humans haven't been around for even one million years – and may not survive long enough to achieve that.

Any suggestion that humans are the 'pinnacle' of evolution would also have to assert that evolution – at least of humans – has stopped. Until humans become extinct, they, like every other organism, will continue to be subject to variation and evolution. As we change our environment, we will in all likelihood change to adapt to it. We have a long period between generations, so our evolution cannot be as rapid as that of insects, bacteria or smaller animals. But it is ongoing nevertheless.

EVOLUTION AT SPEED

We think of evolution as proceeding slowly, but that is not necessarily the case. Although Darwin suggested it was slow and steady, more recent research suggests it can go in sudden rapid leaps. The following organisms have all adapted very rapidly to changes that humans have made to their environments:

- **Fish in the polluted Hudson River have become resistant to toxins which originally poisoned many of them.**
- **An increasing number of elephants are born without tusks, making them safe from ivory-poachers.**
- **Some over-fished species have adapted to reach maturity and breed at a smaller size, making them uneconomic for fishing fleets to catch.**
- **Peppered moths, originally pale coloured, became dark once pollution blackened the surfaces they lived on. (Pale moths were easily visible to predators.) Now, with cleaner air and surfaces, peppered moths have become pale again.**

CHAPTER 2

Why don't we run cars on water?

It costs a lot to run your car, and burning fossil fuels is bad for the environment. Do we have to do it?

How cars work

Conventional cars are powered by an internal combustion engine. The principle of how this works is simple, even though the engine itself looks complicated.

The driving force occurs inside a chamber called a cylinder. Essentially, tiny droplets of fuel are mixed with air, drawn into the cylinder, compressed and set on fire. It's the setting on fire that releases the energy, and this energy is harnessed to move the car along. Here's how it works.

The cylinder is a chamber with strong metal walls. It has to contain explosions, so it must be robust. It has a movable piston, which can go up and down smoothly. The piston fits snugly, and has a good seal with the cylinder – it needs that to maintain pressure and stop the fuel gases leaking out. There are two valves at the top of the cylinder: one to let air and petrol vapour in, and one to let exhaust gases out.

The work is done by fuel, air and a spark plug. The spark plug makes a spark that ignites the fuel. Petrol only burns in air, so the air is as important as the fuel.

The engine takes chemical energy from fuel and releases it as movement (kinetic energy) and heat. The movement of the piston is transferred by a rod to the crankshaft, which turns the piston's up-and-down (linear) motion into round-and-round (rotary) motion. The crankshaft is connected to the drive shaft, which connects to the axles, which turn the wheels.

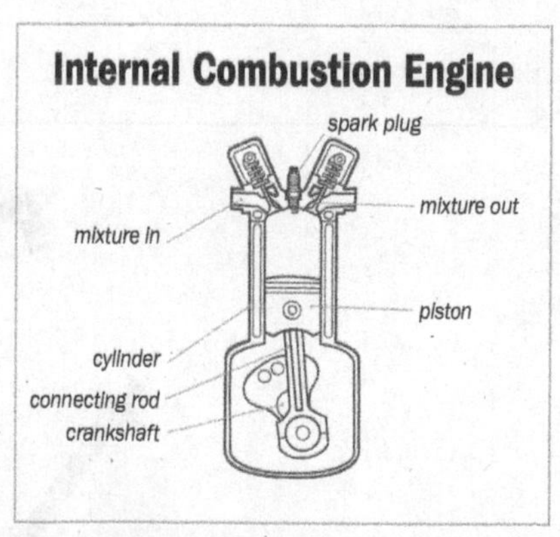

FOUR STROKE ENGINE

The action:

1. Intake stroke – as the piston moves down, air and fuel are drawn into the cylinder

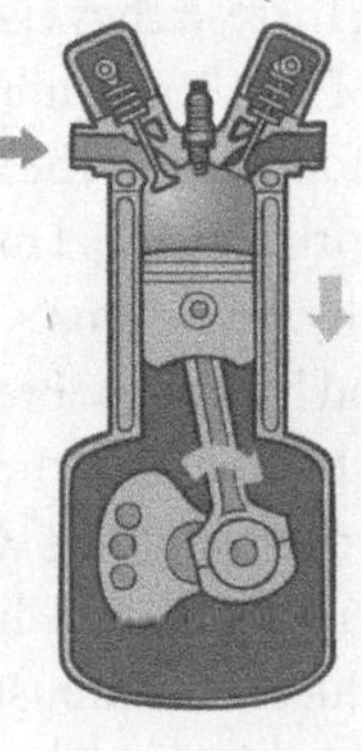

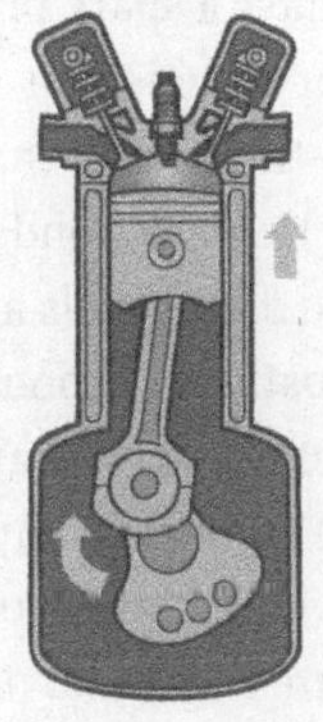

2. Compression stroke – the piston moves up to compress the air-and-fuel mixture

3. Combustion stroke – the spark plug ignites the fuel, forcing the piston down again

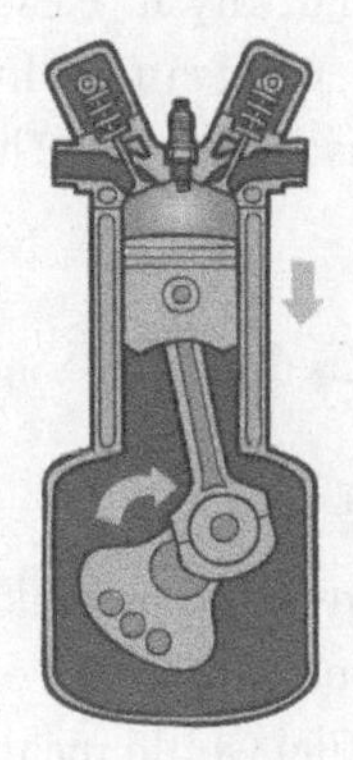

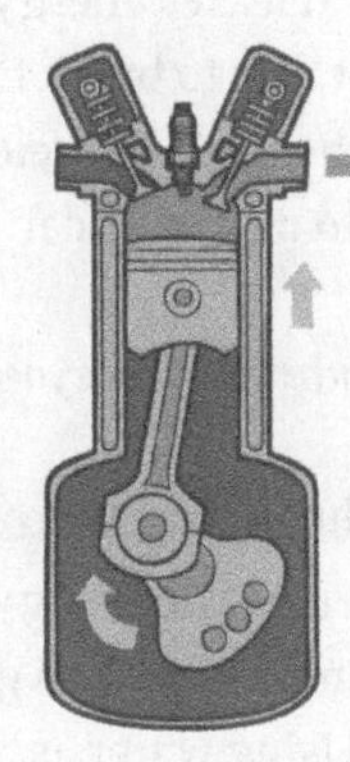

4. Exhaust stroke – the piston moves back up to push out the exhaust gases.

THIS HAPPENS AGAIN AND AGAIN, VERY RAPIDLY.

Energy in and out

If you just mix air and fuel together, they don't burn spontaneously and produce energy. They need to be ignited. In a petrol engine, this is the job of the spark plug. (Diesel engines don't have a spark plug. Instead they use highly compressed hot air to ignite the fuel.) The combustion part takes energy – the energy that comes from the spark. This energy is used to break bonds within the fuel molecules.

All petrol and diesel fuels are hydrocarbons – molecules that consist mostly of carbon and hydrogen. Breaking the bonds separates the carbon and hydrogen in the hydrocarbon molecules. The carbon and hydrogen combine with oxygen from the air to make new molecules – carbon dioxide and water. (These are the waste products of combustion and are pushed out of the exhaust, along with various contaminants and soot.) Making the molecular bonds to form water and carbon dioxide releases energy. Luckily, it releases a lot more energy than it takes to break the bonds in the hydrocarbon molecules, so there is a net release of energy. That's the energy which is used to move the car:

fuel (hydrogen and carbon) + oxygen → water + carbon dioxide + energy

Making and breaking water

The car is driven by the energy released by making the products of combustion. So a good proportion of it comes from *making* water. Using water as a fuel would mean putting in enough energy to break the molecules apart, then using the hydrogen and oxygen to make new molecules in a reaction that produced even more energy than it took to break up the water. Water can be split into hydrogen and oxygen using electricity, in a process known as electrolysis. This is how they produce oxygen on the International Space Station. But it is impractical to power a car this way – you might as well use the electricity directly.

CHEMISTRY AND ENERGY

It takes energy to break chemical bonds, and energy is released by making chemical bonds. A certain amount of energy is associated with each type of chemical bond. The bonds involved in making and breaking apart water are shown on the right. The energy is reported in kJ/mole. A mole is approximately 6×10^{23} atoms or molecules of a substance.

Bond	Bond Energy(kJ/mole)
H-H (hydrogen-hydrogen)	432
O=O (oxygen-oxygen)	494
O-H (oxygen-hydrogen)	460

For a reaction to produce energy, the energy generated by making bonds must be *more* than the energy taken to break bonds.

To make water, the chemical equation is: $2H_2 + O_2 \rightarrow 2H_2O$

This means the energy to break the bonds between hydrogen atoms and between oxygen atoms (on the left-hand side) is:

2 × H-H + O-O 2 × 432kJ + 494kJ = 864 + 494 = 1358kJ

The energy liberated by making bonds between the hydrogen and oxygen atoms to create water is:

2 × 2 × H-O = 4 × HO 4 × 460kJ = 1840kJ

There is a net gain in energy of 1840 – 1358 = 482kJ

If we wanted to use water as fuel, we'd need to supply enough energy to break the water molecules apart (2 H-O bonds at 460kJ each, so 2 × 460 = 920kJ/mole) and find a reaction which would use the hydrogen and oxygen to make bonds producing more energy than this.

Does water burn?

The crucial step in the internal combustion engine is getting energy from burning fuel. Water itself doesn't burn. We can't just put water into an engine and set fire to it. But the principle

GREEN AND CLEAN?

How well a fuel burns depends on the purity of the fuel and getting the right mix of fuel and air – there must be enough oxygen for all the carbon and hydrogen to be converted to water and carbon dioxide. When fuel doesn't burn completely, black particles of carbon are produced and carried by the exhaust gases as soot. Burning diesel fuel accounts for 25 per cent of the soot in the atmosphere – it's really not a clean fuel, because it's rarely very pure.

of making and breaking bonds as a way of getting hold of energy doesn't rely on combustion. It is possible to make a mechanism that uses the energy liberated by making chemical bonds without burning a carbon-based fuel – it just won't look anything like an internal combustion engine. Further, it's not just a matter of finding a reaction that will make bonds. Because this reaction has to happen in a moving vehicle full of people, it has to be capable of being carried out safely and must produce largely non-toxic products. It also has to be cheap. We could get enough energy by passing superheated steam over carbon, but the products would be highly explosive hydrogen gas and highly poisonous carbon monoxide, so it's not a practical idea. We'd also have to carry around coal or some other source of carbon, and have a separate source of energy (such as solar panels) to superheat the water to get steam.

Hydrogen fuel cells

Cars have been developed that run on a hydrogen fuel cell – there are models from Hyundai, Honda and Toyota at the time of writing. Unfortunately, there is only one hydrogen refuelling station in the USA so the car can't go far from home; home must be fairly close to 1515 S River Road in West

Sacramento, California. The car is completely clean, producing only steam as its exhaust. Once condensed, the water is clean enough to drink, according to engineers, but tastes rather flat. Another advantage is that it takes only seconds to fill the hydrogen tank. The dominant competing technology is electric cars, and it takes longer to recharge their batteries, so hydrogen fuel cells win a point there.

The fuel cell itself operates in much the same way as a battery. It works by combining hydrogen with oxygen (easily obtainable from the air) to produce water, the energy of the reaction being harnessed as electricity. All it needs to keep going is a constant supply of hydrogen.

Where does the hydrogen come from?

Hyundai's website rather disingenuously says that hydrogen is a great fuel to use because it's all over the place – 75 per cent of the universe is hydrogen. That's true – but most of it is tied up in things, be it other stars, rocks or human bodies. We can't just catch some hydrogen from the air or space and use it as a fuel.

Currently the hydrogen for hydrogen filling stations comes from natural gas. We could just burn the natural gas, but then the car (rather than the hydrogen manufacturer) would be producing polluting by-products, and these cars sell to customers on their green credentials.

Methane is extracted from the gas, and broken into carbon and hydrogen, with carbon dioxide as a waste product. The

GOING FAR

NASA has used hydrogen fuel cells since the 1970s to launch space shuttles and other spacecraft. The water produced as exhaust is condensed and used as drinking water by the astronauts.

hydrogen is stored in tanks and shipped to the fuel station(s). Researchers are looking for ways of using methane from other sources, such as farm waste, rotting vegetation or cow flatulence. This is important, as natural gas is another fossil fuel which is a limited resource; so using it to provide hydrogen for cars is a finite solution and not as green as the manufacturers would have us believe.

CHAPTER 3

What makes a rainbow?

A rainbow is a beautiful illusion – but how does it come about?

Colour from nothing

A rainbow is created when sunlight falls on droplets of water in the atmosphere. Sunlight is white light, which contains all the colours of the spectrum mixed together. Light is made up of different wavelengths (see *Do we all see the same colours?* page 183), which our eyes see as different colours. The water droplets effectively split the light into its separate wavelengths and send out each wavelength at a slightly different angle so that we see them as a range of coloured strips. You can get the same effect of splitting light by using a glass prism.

IS THERE A POT OF GOLD AT THE END OF THE RAINBOW?

An old legend tells that there is a pot of gold buried at the end of the rainbow. The difficulty is in finding the end of a rainbow in order to dig up the gold. A rainbow is a visual illusion which changes as you move position, so you will never see the rainbow touching the ground. You might see another rainbow further off, or you might find that the rainbow has vanished altogether. This makes it impossible to ascertain whether there actually is gold at the end of the rainbow.

How it works

Light travels in a straight line through a single medium such as air or water. But as it crosses a boundary between two different media, it is refracted (bent). Refraction means that it changes direction slightly from its original path.

This happens because light travels more slowly through a liquid or solid than it does through air. The speed of light is constant in a vacuum, but light slows as it travels through matter. The denser the medium, the longer it takes for light to travel through it. Light travels more slowly through gas than a vacuum, more slowly through liquid than gas, and most slowly through a solid. As the light slows down, its wavelength is shortened proportionally. This splits the white light into a spectrum as different wavelengths are differently slowed. When the light speeds up again, leaving that medium, the refraction is reversed, recombining the white light.

When light travels through a block of glass with parallel sides, such as a window pane, the refraction at the first boundary (air/glass) is compensated for at the second (glass/air) boundary. Therefore, white light is not split into a spectrum by passing through a window.

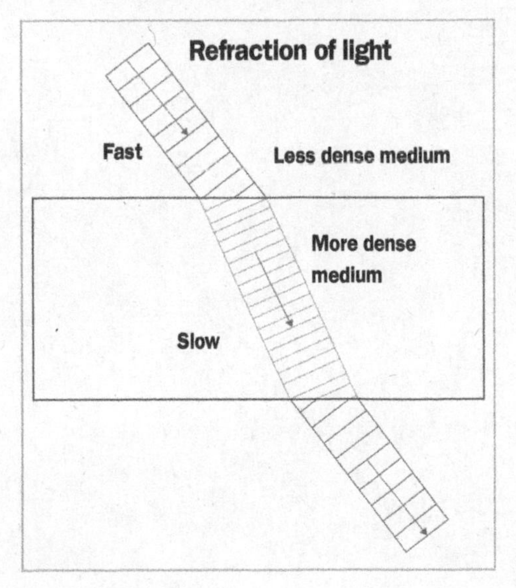

Inside the raindrop

The light gets deflected as it enters the raindrop. But the light from the straw in a glass doesn't split into a spectrum, so clearly something else is going on. As light enters a raindrop, it crosses from air to water and the light is refracted, splitting the colours.

Inside the raindrop, the light travels through the water and meets the water/air boundary the other side. Some light passes straight through, and the colours are recombined into white light. But some is reflected off the inside surface of the raindrop.

On meeting the water/air boundary as it leaves the raindrop, the light is refracted again, further exaggerating the difference in the angle between light of different wavelengths. On leaving the raindrop, red light is travelling at an angle of 42 degrees to the incoming light, and blue light at an angle of nearly 41 degrees. Each raindrop emits light in all parts of the spectrum, but as the light shines out in different directions,

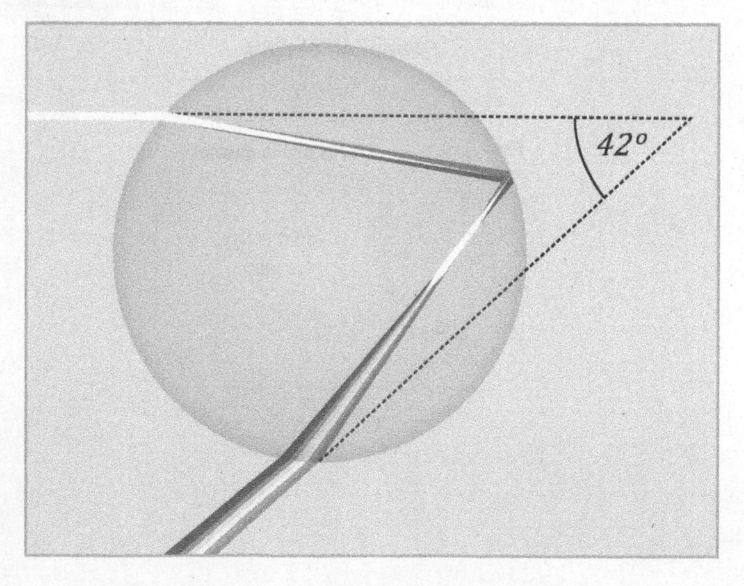

the colour each raindrop appears to show will vary according to where you are standing.

If you look at the red part of the rainbow, all drops in that band are emitting light of all colours, but the red light is coming straight in your direction. In the orange band, the orange light from each raindrop is coming in your direction. Someone in a different place might see this raindrop as blue or green.

The result, with light falling onto a vast number of raindrops, is to create a rainbow. But the rainbow isn't 'anywhere' – it is an illusion or optical effect that seems to hang in the air between the observer and the light source.

Double rainbows

Sometimes the light is reflected off the inside of the raindrops more than once. When this happens you might see a double rainbow. Often the second rainbow is fainter than the first. It also has its colours reversed – they switch round with each reflection.

Finding rainbows

The only place you will see a rainbow is at an angle of 40–42 degrees from the incoming Sun. Imagine a line from the Sun, through your eyes and to the shadow of your head on the ground. The place to look for a rainbow is at a 42-degree angle to this line. The lower the Sun is in the sky, the more vertical the rainbow will appear to be, reaching straight up from the horizon at sunrise or sunset. Conversely, if the Sun is directly overhead, the area at the appropriate angle will be underground, so there will be no rainbow. For this reason, you are most likely to see rainbows in the late afternoon or early morning, but you will never see one at noon. You will see more rainbows if there is more sky; so wide, open country is a better bet than a mountainous area.

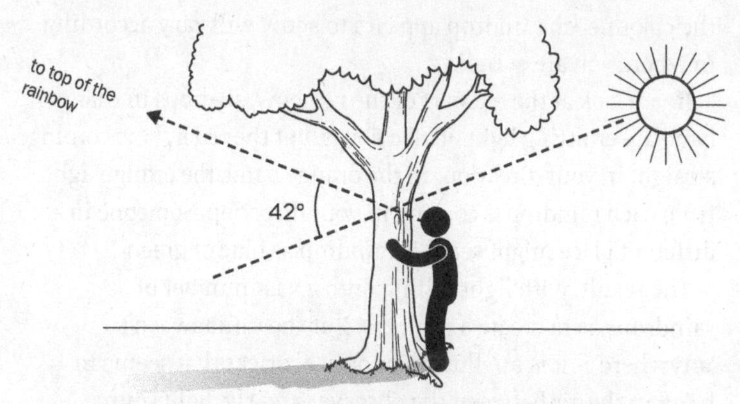

Stand with the Sun behind you to predict the position of a rainbow.

JUST RIGHT

For a rainbow to form, the drops of water have to be just the right size – not too big, and not too small. You can often see a rainbow near a waterfall or fountain, or from a lawn sprinkler, but the water vapour of clouds is too fine to create rainbows. Frozen water – hail or snow – doesn't work either, so we don't get hailbows or snowbows.

Predicting rainbows

You can find out where precisely to look for a rainbow before one appears. This might seem a strange thing to do, but if you want to photograph it or show it to someone, it can be a good trick.

1. After a rainstorm, when the Sun starts to break through, stand with your back to it.
2. Hold your thumb at about a 45-degree angle from your index finger. (Make an L-shape with your thumb and index finger – that's 90 degrees. Now halve the angle by moving your thumb towards your finger.)

3. Hold your arm out and point your finger at the shadow of your head on the ground.
4. Your thumb will point in roughly the direction of the rainbow. You can turn your wrist, keeping your finger and thumb in the same positions, to draw out the full arc of the rainbow.

This isn't foolproof, as a rainbow won't appear if there isn't enough water remaining in the atmosphere or if the Sun is hidden by clouds. But given the right amount of sunlight and water, it should work.

Try taking a photo with a particular object in view, perhaps framed by the rainbow – you can move around and see the rainbow elsewhere. The position of the rainbow is a function of the observer's position – as you move, so will the rainbow. (This also means that no two people see exactly the same rainbow.)

RARE RAINBOWS

Technically, all rainbows should be a full circle, but we seldom see them like this. The exception is when we are looking from a high vantage point, such as an aeroplane, when it is sometimes possible to see an entire, circular rainbow from above.

CHAPTER 4

How does a cat always land on its feet?

It's well known that cats generally land on their feet, no matter how or where they fall. How do they pull off this amazing trick?

Early experiments

The physicist James Clerk Maxwell (1831–79) is said to have experimented with dropping a cat from a height as a student at Trinity College, Cambridge. Maxwell even wrote to his wife, Katherine, excusing himself and claiming that he hardly dropped the cat at all.

> ***'There is a tradition in Trinity that when I was here I discovered a method of throwing a cat so as not to light on its feet, and that I used to throw cats out of windows. I had to explain that the proper object of research was to find how quick the cat would turn round, and that the proper method was to let the cat drop on a table or bed from about two inches, and that even then the cat lights on her feet.'***
>
> James Clerk Maxwell

Maxwell was not the first nor the last person to be interested in how a cat accomplishes this life-saving trick. Cambridge mathematics professor, Sir George Stokes (1819–1903), and Étienne-Jules Marey (1830–1904), French scientist and early cinematographer, were both drawn to it. (As Stokes was professor during Maxwell's time at Cambridge, they might even have shared cat-dropping notes.) Marey filmed a cat falling and righting itself, using the individual frames to see exactly what happens. His photographs were printed in *Nature* in 1894. If you look carefully at them and read the account of what the cat does, all is revealed.

Learning to fall

A cat dropped upside down, from any height greater than about 30cm (12in), is able to right itself before it reaches the ground, so it almost always lands on its feet. This self-righting reflex emerges in kittens at 3–4 weeks old and is fully in place by 6–7 weeks. Even a cat without a tail can right itself – the tail is not necessary to the manoeuvre, though it can be used as an option.

How the cat does it

So how does the cat achieve a feat that no other animal is known to be capable of?

The cat begins by working out which way is down. This might seem fairly obvious, but it is a vital first step – its legs need to be pointing downwards to make a successful landing. It can do this by looking, but balance organs in the inner ear help, too. Then it needs to twist so that its feet are underneath. The cat can perform the following three steps only because it has a very flexible spine. In addition, its clavicle (collarbone) is free-floating and therefore effectively useless. (This gives the cat the added advantage of being able to move its shoulders far more freely than a human when trying to fit through a small space – a cat can squeeze its shoulders through any space its head will fit through.)

Firstly, the cat bends in the middle to form something like a boomerang shape. As a consequence, the axes of rotation for the front and rear halves of the body are different. This is important, as it means the cat can preserve angular momentum (in effect, its spinning force) as it twists.

The cat tucks in its front legs and extends its back legs so that the front end rotates faster than the back end. Think of how a figure skater holds her arms close to her body to spin quickly and holds them out to rotate slowly – the cat is doing the same thing, but with different halves of its body. The front half rotates through a much larger angle than the back half because the front paws are held close to the body. It can rotate as much as 90 degrees. The back half, with extended paws, rotates very little – maybe only 10 degrees. So if the cat's original rotation was zero degrees, the front is now at 90 degrees and the back at 10 degrees.

Next, it switches things round. It tucks in its front legs and extends its rear legs so that its back half rotates through a larger angle than its front half. After this manoeuvre, the front and back are both at 10+90=100 degrees. At this point, it would land on its side, which is no good. So it does the whole thing again (it doesn't have to do the greatest turn it can each time).

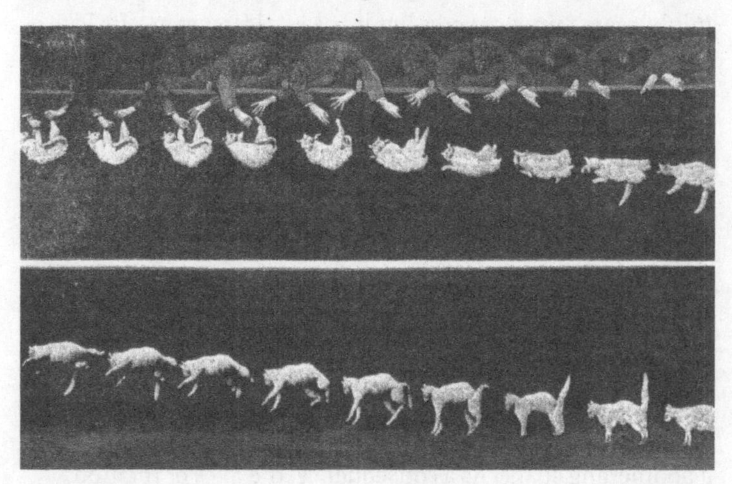

The cat can acquire angular momentum as it falls because it can bend and use the two halves of its body separately. A

HUMANE RESEARCH

Dropping cats from a great height would be a tricky experiment to justify – even Maxwell felt the need to defend himself, long before the existence of ethics committees. Luckily, in 1987, someone thought of a more humane method. They studied the injuries sustained by cats brought into the New York Animal Medical Center after falling from high buildings. From 132 feline patients, it emerged that injuries increased in severity up to a seven-storey drop and then decreased. The researchers suggested that the cats reached terminal velocity at seven storeys and then relaxed. The impact with the ground was less damaging because of their relaxed state.

However, as one critic pointed out, the results are flawed as they don't take account of fatalities: no one takes a dead cat to the emergency clinic, so the data are incomplete and the conclusion unsound. There is still something interesting, though, in the reducing severity of the injuries – and the fact that cats survived at all. Not many people would be in the emergency department after a fall of seven storeys.

rigid body couldn't do that – both ends would have to rotate together. Mathematical models of the cat self-righting treat it as two cylinders, one for the front half and one for the back, and were not completed until the late 20th century. But the cat can do it, whether or not we can describe it mathematically.

FAST, BUT NO FASTER

Terminal velocity is the maximum speed an object of a specific mass can achieve falling under Earth's gravity and in the atmosphere. Initially, a falling object accelerates at 9.8m (32.2ft) per second per second (m/s^2). After a certain point – terminal velocity – the pressure of the air beneath the falling object prevents it accelerating any further and its speed remains constant until it lands. The terminal velocity of an object depends on its mass and its shape, as air resistance (drag) is a major component. Cats reach terminal velocity at around 100km/h (62mph), whereas for a human terminal velocity is 210km/h (130mph) in the freefall position used by parachutists. A cat hitting a pavement after falling from a high building would be travelling at only half the speed of a person in the same circumstances. Furthermore, the cat weighs less than a person, so the shock to its body is less. If a person fell seven storeys, even landing on their feet would not be good for them, but as the cat is considerably lighter it will hit the ground with less force.

CHAPTER 5

Why is soil brown?

We're all used to children asking why the sky is blue. But what about something more down-to-earth?

What is soil?

Soil is more or less everywhere – in our gardens and fields, beneath roads and buildings. There's a form of soil, called sediment, at the bottom of rivers, lakes and ponds. It's only absent in places where there is sand or bare rock instead of soil.

Soil has a different composition in different places. It can be sandy, heavy with clay, or rich with humus (composted matter). It's usually a shade of brown – reddish-brown or so dark that it's nearly black or a pale, sandy brown, for example.

Just as the layer of air above us is called the atmosphere, the soil layer has a name, too – it's called the 'pedosphere'. Soil comprises bits of mineral (rock) and organic (plant and animal) matter. It's loosely jumbled together, and as the bits are odd shapes and do not fit together neatly, there are gaps between them. The gaps make soil porous, so it can hold liquids and gases. This means that soil has the three states (or 'phases') of matter – solid, liquid and gas – all mixed together.

Dead things

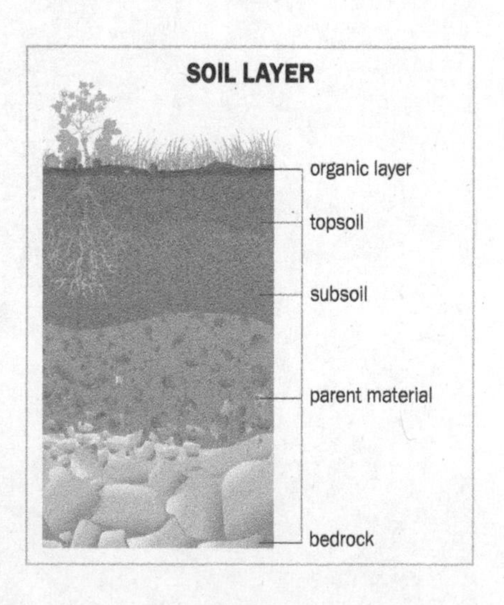

Organic matter gives soil its brown colour. Leaves and other plant debris fall to the ground, as does animal waste, including faeces, fur, and so on. All these are acted on by the many microbes in the soil, which break them down. It's sometimes a slow process, especially when it comes to breaking down bones, which takes years. Decomposition occurs when the microbes release enzymes that break chemical bonds within the organic matter.

Although the microbes absorb many of the chemicals produced by decomposition (it's their source of food), there tends to be a surplus of carbon. In time, the microbes die and are themselves subject to decomposition in the same way. Again, most of the chemicals are absorbed by other microbes, but not all – there is some inefficiency in the system. The result is that there is surplus carbon lying around, and as carbon reflects light, it looks dark-coloured or brown.

The leftovers from microbial activity form humus in the soil. Unlike plant matter, humus doesn't have a cellular structure, but is more amorphous – gungy, in fact. Humus can remain in the same state for thousands of years. Usually there are microbes and bits of identifiable remains mixed into it, which show up under microscopic examination. But real, pure humus is organic gunge.

What about the rock?

The mineral component of soil is made up of tiny fragments of rock that have been broken up by erosion, then carried by water, wind or ice to wherever they end up. Typical minerals in soil are quartz (silicon oxide), calcite (calcium carbonate), feldspar (a compound of potassium, aluminium, silicon and oxygen) and mica (also called biotite, a compound containing potassium, magnesium, iron, aluminium, silicon and oxygen). Most of these tend to be white, grey or brownish, although impurities can give them extra colour.

Usually soil is brown because of the carbon present in the humus, but if the soil is particularly impoverished and mostly mineral, it takes the colour of the dominant mineral. The soil in Hawaii has a reddish tinge caused by its iron content – rust,

HOW OLD IS SOIL?

Most soil on Earth dates from the Pleistocene epoch, 2,588,000 to 11,700 years ago. No current soil is more than 65 million years old, but fossilized soil dates back much further, even to the earliest time of Earth's existence, the Archaean eon. That time period extended from the Earth's origins until 2.5 billion years ago. As there were no plants and animals then, just microbes, the soil would have been very different.

MICROBES GALORE

There are thought to be up to a billion cells in every gram of soil, possibly representing anything from 50,000 to a million different species of microbe. Most of these species have not yet been described.

which is iron combined with oxygen, is red. Geologists call substances 'soil' which many of us would call sand. So the deserts might have 'soil' that looks white, black (volcanic sand), yellow, red or even green (olivine sand, found in Hawaii). Occurring mostly in places with relatively little vegetation and animal life, these soils have a high mineral content and very little humus.

And why is the sky blue?

As light reaches Earth's atmosphere, it bounces off gas molecules and scatters. Light from the Sun is white – or, rather, it is all the colours of the rainbow mixed together so that it looks white. Because blue light has the shortest wavelength, it is scattered more than the light of other colours. On a cloudless day, you'll see that the sky is bluer overhead than it is at the horizon. This is because light at the horizon has been scattered and rescattered by more air. It has bounced around so much that some comes back to us remixed into white light.

At sunrise and sunset, the Sun is low in the sky and the light has to pass through more of the atmosphere to reach your eyes (because it's further to the horizon than to the edge of the atmosphere looking straight up).

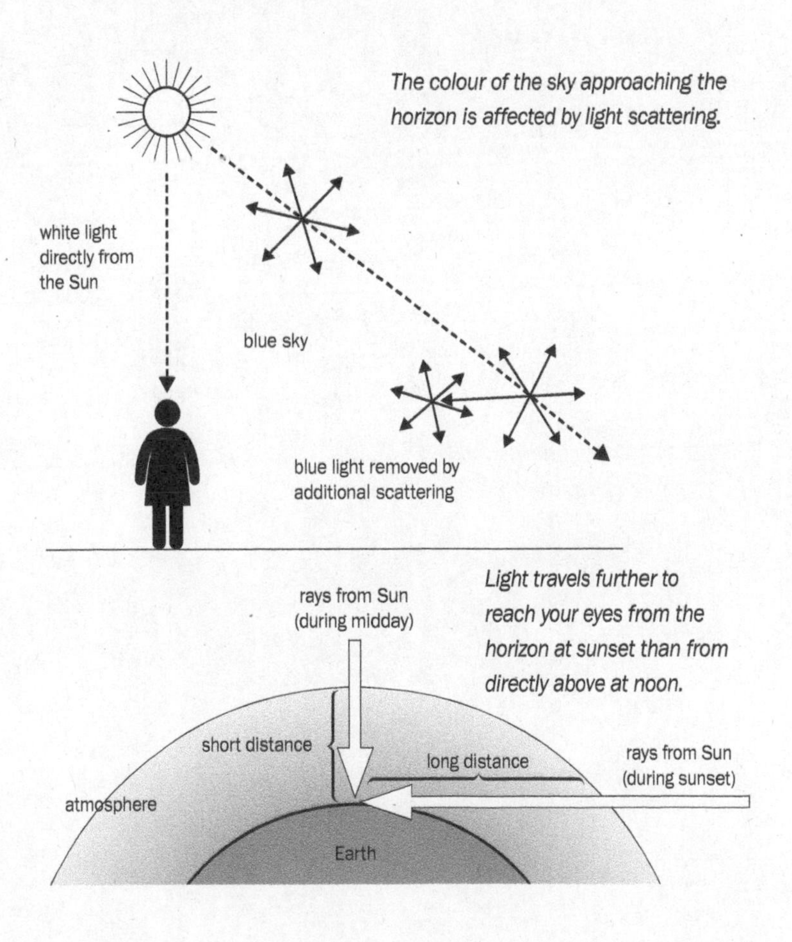

The colour of the sky approaching the horizon is affected by light scattering.

Light travels further to reach your eyes from the horizon at sunset than from directly above at noon.

More and more yellow, and even red light, can reach you. If the air is heavy with dust particles or liquid, the effect is even greater, which is why pollution and volcanic eruptions can produce brilliant sunsets.

CHAPTER 6

Why don't we go to Mars?

It's more than 40 years since men last walked on the Moon. Is a trip to Mars the next obvious step for humankind?

What's stopping us?

There are several problems with getting humans to Mars, though NASA and some independent commercial organizations are hoping to solve them and send a crew in the 2020s or 2030s. Some of the tricky issues are as follows:

- Mars is a long way off
- Keeping humans alive, healthy and sane in space for the duration of a trip to and from Mars is a challenge
- Carrying the fuel and supplies needed for the trip is beyond current capabilities
- Landing a spacecraft on the surface of Mars is very tricky, as is taking off again later.

ARE WE NEARLY THERE YET?

The distance from Earth to the Moon is 384,400km (238,855 miles). Driving to the Moon in your car would take 160 days at 100 kph (62 mph), 24 hours a day. To drive to Mars, at its closest, would take 23,360 days (nearly 64 years).

Far, far away

Distances in the solar system are vast. Images like the one below show us that Jupiter is larger than Earth and Neptune is

furthest away, but they don't give us an accurate impression of the distances between planets or the true disparity in size.

Mars and Earth are on different orbits around the Sun. It takes Mars 687 Earth days to orbit the Sun and it takes Earth 365.25 days – their orbits are not synchronized. This means that sometimes they are closer together and other times they are farther apart. At their closest, there is 'only' 54.6 million km (33.9 million miles) between them. But when they are at their most distant they are around 400 million km (248 million miles) apart. The average distance between Earth and Mars is around 225 million km (140 million miles).

The points at which Mars and Earth are closest do not come round on a regular pattern. In August 2003, Earth and Mars were closer than they had been in nearly 60,000 years. They'll be just as close again on 28 August 2287 – a gap of just 284 years this time. The proximity of the two is affected not only by their orbits, but also by the effects of gravity exerted by other planets.

If we were to plan an expedition to Mars, we would need to make the journey there and back as short as possible. It wouldn't be sensible to set off at a time when Earth and Mars

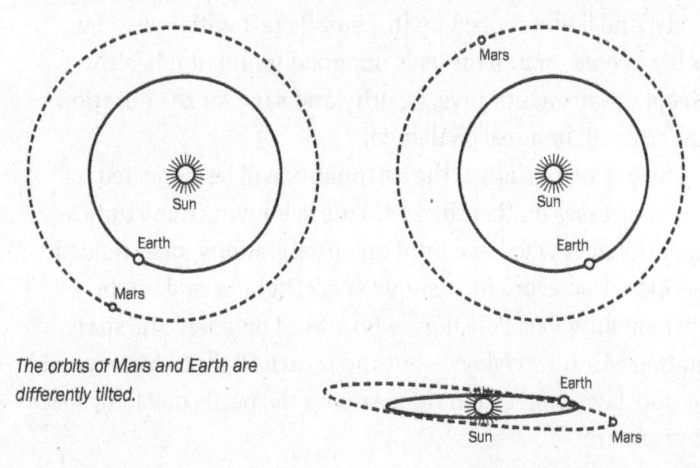

The orbits of Mars and Earth are differently tilted.

were 400 million km (248 million miles) apart. And then there's the return journey. If we pick the shortest time to get to Mars, it's going to take longer to get back, as the planets will have moved further apart. How far they will have moved depends on the length of time it took to get there and how long we stay on Mars before heading home.

A trip to Mars won't require the spacecraft to travel in a straight line from Earth; instead it will enter an orbit around the Sun that will eventually bring it into conjunction with Mars. The route map (see diagram opposite) is optimized for economy. The potential launch date for this trajectory occurs every 26 months.

Using current rocket technology, it's going to take nine months to reach Mars and nine months for the return trip. The astronauts will also have to stay on Mars (or in orbit around it) for three to four months, until Earth is in the right place for the return journey. That makes a total trip of at least 21 months.

Stresses and strains

Being in space, in zero-gravity, is not very good for the human body. And being cooped up in a small craft with just a few other people for 21 months is not good for mental health. Keeping astronauts alive, healthy and sane for the duration of the trip will be a real challenge.

During space flights, the astronauts will be subjected to many stresses on their bodies. This is known from studies of astronauts who have lived on space stations for extended periods. The record for a single spaceflight is held by the cosmonaut Valeri Polyakov, who stayed on board the space station Mir for 437 days – but the return flight to Mars would be 200 days longer than that. Among the perils of a long space journey are:

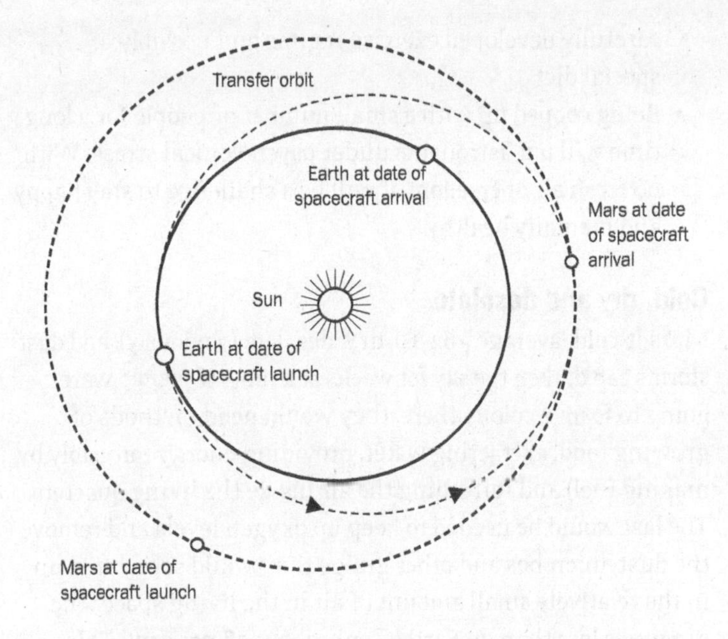

- Bombardment with protons from solar flares, gamma rays from newborn black holes and cosmic rays from exploding stars. A trip to the Moon is largely sheltered from these hazards as the perils are deflected by the Earth. To help protect astronauts, NASA is investigating ways to build a spacecraft from plastics, carbon nanotubes, or even wrapping a liquid hydrogen fuel tank around it.
- Lower gravity and micro-gravity – our bodies are used to working against gravity and without it our muscles quickly waste away (atrophy) and bones become less dense and more prone to breaking. As the heart is made of muscle, a weakened heart will be a problem for astronauts returning to Earth. Astronauts might also be too weak to carry out physical labour when they arrive on Mars. On space stations, astronauts take regular exercise to combat these problems. On a much longer trip, they would need a

carefully developed exercise regime and possibly a special diet.

- Being cooped up with a small number of people for a long time will put astronauts under psychological stress. With no fresh air or freedom, it will be a challenge to stay happy and mentally healthy.

Cold, dry and desolate

Mars is cold (average –62 °C), dry, desolate (obviously), and dust storms can darken the sky for weeks at a time. If people were going to form a colony there, they would need methods of growing food, extracting water, providing energy (probably by making fuel) and refreshing the air inside the living quarters. The last would be needed to keep up oxygen levels and remove the dust, microbes and other gunge that would soon build up in the relatively small amount of air in the living space. The gravity is less than on Earth – only about 38 per cent. This means that someone who weighs 100kg (220lb) on Earth will weigh only 38kg (84lb) on Mars. Bones and muscles need to do less work and will weaken. The atmospheric pressure is around 1 per cent of that on Earth. Colonists will be wearing space helmets and air tanks all the time they are outside.

Big, big spacecraft

A spacecraft carrying a human crew to Mars will need to be much larger than those which have carried robotic rovers in the past. It will need to carry the equipment and crew for the mission, plus supplies for the crew, and fuel. A human crew needs a lot of inconvenient weighty things such as food, water and medical supplies if they are to survive for 21 months away from Earth. NASA estimates that the supplies needed for a crew of six would weigh in the region of 1.4 million kg (3 million lb). The space shuttle can lift about 22,500kg (50,000lb)

WHO WOULD GO?

You'd think, given all the inconveniences, from muscle wastage to death, it might be hard to find volunteers to go to Mars. Not at all. The not-for-profit organization Mars One, based in the Netherlands, began recruiting its first batch of astronauts in 2013. It plans to land the first group of four on Mars in 2023, and it's a one-way ticket. They will send supplies in advance, and the colonists will use them to build their colony when they get there. Then they will stay there forever, never returning to Earth. Further batches of settlers will arrive every two years. Mars One had more than 100,000 applicants competing for the first four places.

into space, so it would take around 60 launches to get all the equipment to build and stock the spacecraft. Even the lander would have to be at least ten times the mass of landers used in the past on Mars.

All that fuel

The problem with heavy things is that they take more energy, and therefore more fuel, to move ($F = m \times a$, force = mass × acceleration). Most of the fuel is used to lift off – it takes a lot of energy to escape gravity each time. It also takes some fuel to slow down on the approach to each of the planets, adjusting the speed so that the craft can be captured by gravity and so it can land safely on the surface.

Fuel to take off from Earth is not much of a problem as it only has to be transported to the launch pad. Typically, a spacecraft launched from Earth burns fuel then discards the modules that carried it, leaving a much smaller vehicle to travel on into space. The rest of the fuel is far more problematic as the craft has to carry it. This means that the spacecraft has to be even bigger because it has to carry more fuel, and then it needs to be even bigger again, so needs *even more fuel than that*. . . and so on.

Spacecraft use relatively little fuel to travel through space. As there is no air resistance, there is nothing to slow them down once they have started moving, so momentum keeps them going in the same direction. As the spacecraft approaches Mars, it will need to slow down sufficiently to be captured by the gravitational pull of Mars and be drawn into orbit.

Traditionally craft use retro-firing rockets, which work to push the craft backwards when they need to slow down. But this means carrying more fuel for those rockets. NASA could use a technique called 'aero-capture', which involves plunging the craft down towards Mars to skim the atmosphere and then relying on friction with the atmosphere (drag) to slow it down. The trajectory has to be carefully calculated, however. If the craft goes in too far, the heat from the friction will burn it up, even with sophisticated heat shielding. If it doesn't go in far enough, it will not slow down sufficiently and will whizz past the planet. Finally, the craft has to return from Mars to Earth. That involves lifting the lander off the surface to rejoin the main craft in orbit and then pulling the main craft away from Mars and towards Earth.

There and back again

Landing on Mars is tricky. There have been plenty of mishaps in the past; around two-thirds of the 40 or so missions to Mars have failed at some point between launch and touching down on the planet. The very first mission, Mars 2, launched from the USSR, crashed in a dust-storm on landing. Although the success rate has improved, landing is still far from foolproof. It seems likely that a large part of the craft would remain in orbit over Mars and a smaller lander would be sent to the surface. The orbiting module could provide back-up and a safe haven – and might even be able to drop supplies or launch an emergency rescue mission, if necessary.

The size of the lander needed will be substantial – unlike anything so far landed on Mars – and that makes landing safely an extra challenge. Previous landers have either had legs or airbags. The lander descends to within a few metres of the surface, using retro-rockets to slow its descent to zero, then free-falls the last bit. One way would be to use a 'sky crane' – a mechanism that would lower the lander gently to the surface at the point where the retro-rockets exactly counteract gravity (when the craft is stationary, but above the surface). A similar system was used to land the Mars rover *Curiosity*. It remains important to pick exactly the right landing site and time – large rocks, too much of a slope, or a major dust-storm could scupper any type of landing.

Design for living pods produced by Mars One, the commercial organization planning a colony on Mars.

So far, no rock or soil samples have been returned to Earth from Mars because of the difficulty of launching a vehicle from the surface of a distant planet. None of the infrastructure that exists on Earth to launch a craft is in place on Mars. The gravity there is around four times the gravity of the Moon, so it is considerably more of a challenge than relaunching the *Apollo* lander. The size of the lander, coupled with the gravity and atmosphere of Mars, means that a good deal more fuel will be needed to lift off from the surface of Mars than from the Moon.

It looks possible that, as British Astronomer Royal Sir Martin Rees has said, the first Mars missions might be a one-way trip – just as Mars One intends them to be.

LOST ON MARS?

In the film *The Martian* (2015), Matt Damon plays the part of an astronaut who has been left for dead on the surface of Mars. When he revives, he struggles to survive with the meagre supplies he has been given while his space-colleagues and Mission Control on Earth attempt to launch a rescue mission. Given how many years it takes to plan a space mission, the prospect of a speedy rescue seems remote.

The film was screened on the International Space Station on 19 September 2015.

WHY NOT SEND A BOT?

Why don't we build super-good robots and send them to Mars instead of humans? After all, machines are better able to cope with the conditions. But human explorers can still do lots that a machine can't (see *Could intelligent machines take over?* page 199), so the best solution is to send humans with smart machines. This means the human craving for adventure can be satisfied, with technology doing the hard work.

CHAPTER 7

Could we bring dinosaurs back to life?

The central premise of Michael Crichton's 1993 novel *Jurassic Park* (and the many films based on it) is that we can recreate dinosaurs from remnants of their DNA.

In the novel, Jurassic Park was a theme park that had been built on an island and populated with dinosaurs cloned from dinosaur DNA. The DNA was extracted from the guts of mosquitoes which had sucked blood from dinosaurs; the insects had been trapped in pine resin which had solidified into amber over time. More than 65 million years later, scientists extracted the DNA and implanted it in a living egg. Is this feasible?

Dead but not gone

When an organism dies, whether it is a plant or an animal, it begins to decay – to be broken down by microbes and the action of chemicals and weather. Before this it is theoretically possible to take tissue samples and create a clone of the organism from its DNA (see box opposite and *What's the difference between a person and a lettuce?* page 207). This is how services offering to recreate a beloved pet dog or cat work. They take a cell from the dead pet and add its nucleus – which contains the DNA – to an egg cell from another animal of the same species. The nucleus has been removed from this host egg cell, so all the DNA of the new organism comes from the original pet. The egg cell develops into an embryo that is an exact copy (a clone) of the pet, which is implanted into a surrogate mother of the same species. The first mammal cloned in this way was Dolly the sheep, created at the Roslin Institute in Edinburgh, Scotland, in 1996.

As bodies decay, it becomes harder to extract intact DNA from their cells. Some tissues decay faster than others, and as almost all cells contain DNA, it is often possible to extract DNA quite a long time after death, if the whole body is available. DNA lasts much longer if the body is frozen. It doesn't have to be deliberately frozen in laboratory conditions. Mammoth DNA has been recovered from the shrivelled remains of woolly mammoths preserved in the Siberian permafrost. This is useful for scientists who want to compare the genetic

DNA

With the exception of some viruses which are on the borderline between living and non-living things, all organisms have genetic material called DNA (deoxyribonucleic acid). DNA exists as long, complex molecules that carry a code in the form of the arrangement of molecular groups called 'bases'. There are four bases: cytosine, guanine, adenine and thymine. They always occur in pairs, and these pairs are always consistent: each cytosine is always paired with guanine and adenine with thymine. Almost every cell in the body contains a copy of the organism's DNA arranged into genes along long strands of genetic material called chromosomes. The unique structure of the DNA defines that individual organism. All organisms of the same species have equivalent genes and chromosomes, but slight variations in the exact sequence of bases in the DNA account for any differences between individual organisms.

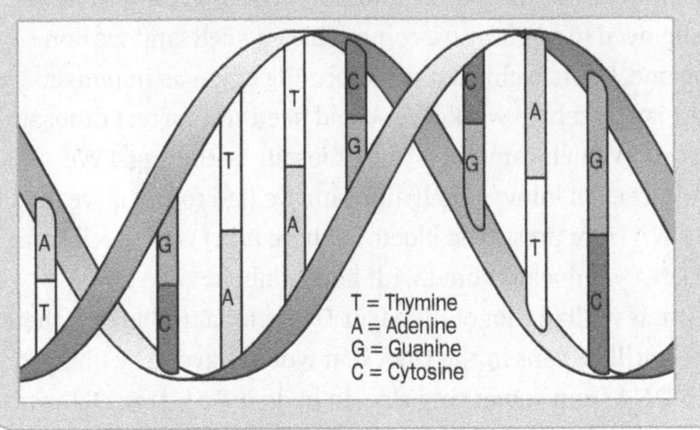

make-up of mammoths with their nearest living relatives, the elephants. Some scientists even hope that one day it could be possible to recreate mammoths. This could perhaps be achieved by using DNA from frozen remains and an elephant egg cell from which the nucleus has been removed – just like cloning a favourite dead pet.

So far, no attempt to bring back extinct animals by cloning has been successful. An extinct Iberian lynx, cloned in 2009, died soon after birth because of lung defects. But endangered animals have been successfully cloned. So in theory it should be possible to bring back extinct animals like the mammoth, if the quality of DNA is good enough.

What about dinosaurs?

The *Jurassic Park* scenario is not plausible, sadly. DNA degrades quite rapidly (in thousands of years), and would not survive the 65 million years that have passed since the last non-avian dinosaurs died, even if it were encased in amber. A research team at Murdoch University in Western Australia set an upper limit for the survival of DNA to 6.3 million years – only a tenth of the way back to the dinosaurs.

Even if we did find some dinosaur DNA to grow a clone we would need to put it into a compatible egg cell, and we don't have one. Using a chicken or a crocodile egg – as in *Jurassic Park* - isn't likely to work. We would need the correct dinosaur egg cell. (Which came first, the dinosaur or the egg?) We wouldn't even know which dinosaur we had found if we took the DNA from preserved blood, as there is no way to tell from the DNA – it doesn't come with handy labels.

Even if we had a lot of dinosaur DNA and a usable egg, there would still be gaps in the DNA that would have to be plugged with DNA from something else. In *Jurassic Park*, frog DNA was used to fill the spaces – no genes for dinosaur intestines were present. Would this dinosaur-frog chimera still 'count' as a dinosaur?

Build your own

If the recovered-DNA route is not going to work, is there another way to bring back dinosaurs? It seems there might be a way to recreate them, which is not exactly bringing them back but still gives us a dinosaur of sorts. Paleontologist Jack Horner has proposed that we could use reverse genetic engineering to recreate a dinosaur by changing the genetic make-up of birds. This would involve incrementally adding dinosaurian features to birds, effectively stepping back through evolution. It wouldn't produce a dinosaur of a type that actually existed, but might, if it worked, produce something like a dinosaur.

Birds from dinosaurs

It's often said that birds evolved from dinosaurs around 150 million years ago but, strictly speaking, birds are avian dinosaurs and the extinct dinosaurs are non-avian dinosaurs. In other words, birds *are* dinosaurs, just not the type we usually think of when we see the word dinosaur. Birds are a development of theropod dinosaurs – the types, like Tyrannosaurus rex, that walked on strong hind legs and were typically carnivorous. Like birds, many dinosaurs probably had feathers, though not necessarily full plumage. Some might have had quills or proto-feathers, and even T. rex might have had fluffy down as a baby (or chick). All dinosaurs laid eggs, as

PLEISTOCENE PARK

Pleistocene Park is a reserve in Siberia where the Russian scientist Sergey Zimov is running a project to try to recreate the subarctic steppe grassland of the last Ice Age. Zimov's theory is that over-hunting rather than climate change wiped out the large Siberian mammals, including mammoths. He suggests that if they can be reintroduced, the grasslands will help to retain the permafrost and cut the escape of methane from melting permafrost that contributes to climate change. He hopes to recreate mammoths by a process of de-extinction.

birds do. Non-avian dinosaurs had beaks or snouts with teeth, tails with bones, separate digits on their forelimbs (fingers, effectively) and claws on those digits. Most birds living today have none of those features. The only known exception is the hoatzin bird of South America, whose chicks have claws on two of their wing digits. The hoatzin (or stinkbird) is thought to be the last survivor of a line that died out 64 million years ago, just one million years after the extinction of the non-avian dinosaurs.

The earliest known bird-like dinosaur is Archaeopteryx, which lived 150 million years ago. It had feathers, a bird-like shape, and wings. But it also had teeth in its beak, claws on its wings and a long tail-bone beneath the feathers. All this is clear from some remarkably well-preserved fossils.

The puzzle for those who want to recreate a dinosaur is to find out which bits of DNA coded for the snout, teeth, tail bone, separate fingers, claws and so on. To build a dinosaur that's really not closely related to birds at all – such as a triceratops, diplodocus or ankylosaurus – would be pretty much impossible if starting solely from birds.

Dinosaurs from birds

Horner's plan sounds mad, and many other paleontologists would agree with that verdict. But in 2015 a team working at Yale and Harvard claimed they had managed to interrupt the action of proteins that lead a bird to develop a beak and had forced a chicken embryo to grow a snout instead.

A bird's beak is formed from two patches of cells in the embryo that extend and grow into hardened tissue jutting out of the front of the mouth. In other animals, the same patches exist but are smaller and grow differently, becoming part of the jawbone. The front teeth are rooted in that piece of bone. Using a chemical to interrupt the action of proteins in the developing

DINOS WITHIN INSECTS WITHIN TREES

The conceit in *Jurassic Park* is that dinosaur DNA is retrieved from blood found in the guts of insects that fed on them. The insects had been trapped in amber, a resin exuded by trees which then hardens. In fact, if dinosaur DNA were preserved in blood in the gut of an insect, it would be very difficult to disentangle the dinosaur DNA from the insect DNA, and the DNA of any other meals the insect had recently eaten. Without any dinosaur DNA to match, scientists wouldn't know which fragments went togethor and which belonged to another organism.

embryo, researchers were able to stop the development of a beak and make the cells develop into a jawbone instead.

Other paleontologists who dream of recreating a dinosaur have worked on extending the tail bone of chickens, reverting feathers to scales and adding teeth to the beak. One enthusiast is paleontologist Hans Larsson of McGill University in Canada. He is convinced that many dinosaur genes are preserved in bird DNA and could be turned back on, expressing dinosaur features such as teeth and tail bones. He notes that a chicken embryo has 16 vertebrae in its tail early on, but by the time the chick hatches it has only five. Turning off the gene that re-absorbs the tail would lead to a chick with a tail – another step towards dinosaurhood.

'What we're doing is trying to identify the historical pathway that evolution took to get from dinosaurs to birds. Then we can just reverse that and go back to having an animal that looks kind of like a dinosaur.'

Jack Horner,
Professor of Paleontology,
Montana State University

Dino or not-dino?

Not all experts agree that developing a tail bone or jaw with teeth are genuine steps in de-evolving birds, but if they are, and if similar steps can interfere with other areas of development, it might be possible to grow a pseudo-dinosaur. Even so, would it count as a dinosaur? As Horner has pointed out, it could be possible to engineer dinosaurs to our own design – a micro-stegosaurus as a pet, for example. But would it be a dinosaur if it didn't have dinosaur DNA? Biologically, no – as Horner acknowledges. A chicken that looks like a dinosaur (chickenosaurus, as Horner calls it) would still be a chicken, just a modified one. But to visitors to any future real-world 'Jurassic Park', it might well be good enough.

> **EVOLUTION IN THE EMBRYO**
>
> **The embryos of many organisms show features that have been genetically turned off but might have been fully expressed in ancestral forms. Early observation of this led to the erroneous belief that animals race through all the stages of their evolution during their embryonic development.**

CHAPTER 8

Could a super-volcano kill us all?

Volcanoes may lie dormant for hundreds of years, but the longer they do so, the more devastating the results may be when they finally erupt.

Every so often, someone suggests that a super-volcano eruption could be imminent and is about to kill us all. It's a great scenario for disaster movies – but how realistic is it?

From little volcanoes to super-volcanoes

Volcanoes are not all the same. Some erupt little and often, causing relatively light damage. Some erupt more violently but at longer intervals, often causing considerable damage. And some lie dormant for centuries or even millennia, and then erupt with catastrophic consequences. Some of the most destructive of all are the super-volcanoes, which are bigger and more powerful than any of the standard types of volcano. They are also *so* large that they are difficult to spot. Not just a single large mountain or even a wide shield, they often show just as a dip in the ground. A vast caldera (volcanic crater) many kilometres across can seem like a calm lake or peaceful, fertile valley, but in fact it's treacherous and deadly, just biding its time.

Magma builds up under a super-volcano for thousands of years. It is so hot that it melts the crust around it, adding to the volume and pressure of magma. When the eruption finally comes, it is powered by such an immense volume of magma under pressure that the effects are truly devastating. When it's all over, the land above collapses into the space left by the emptied magma chamber, creating a huge basin – the caldera. There has not been a super-volcano eruption in recorded history, but geological evidence shows us what it could be like. And there are the calderas left by previous eruptions which show us where super-volcanoes once were – or still are.

Once, long ago . . .

The last super-volcano to erupt was Toba in Indonesia (VEI 8 – see box opposite). It left a caldera that has become Lake

FROM SMALL VOLCANOES TO SUPER-VOLCANOES

Not all volcanoes look like mountains. Some of the most deadly don't look like a 'typical' volcano at all.

- **Volcanic cones** or **cinder cones** are cone-shaped and generally fairly small, at 30–400 m (100–1,300 ft) high. Sometimes they appear on the sides of larger volcanoes and usually erupt only once.
- **Shield volcanoes** are large, shallow mounds the shape of a shield. They form when lava erupts slowly and runs over the ground for some distance before hardening. They build up gradually over many years or even centuries.
- **Stratovolcanoes** are mountain-shaped, built up of the lava and ash from repeated eruptions. They can erupt explosively, producing devastating floods of lava and hurling flying lumps of semi-molten rock high into the air. They produce scorching winds that travel at hundreds of kilometres an hour and vast, choking clouds of ash and dust. Vesuvius, in Italy, which destroyed the Roman city of Pompeii, is a stratovolcano.

TYPES OF VOLCANO

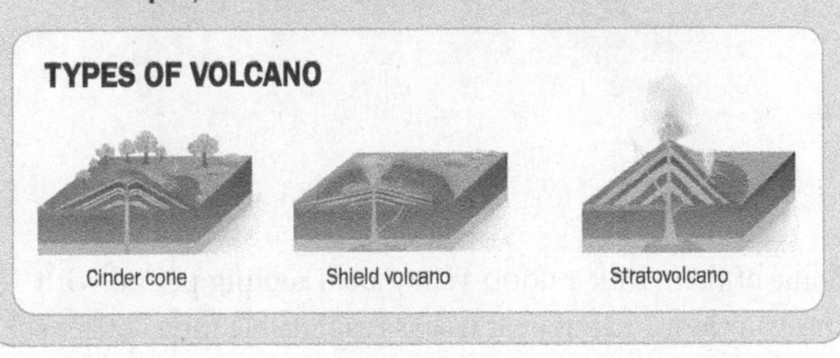

Toba, 100 km (62 miles) long and 50 km (31 miles) wide. That eruption was around 75,000 years ago, although evidence of it was only discovered – and with it the existence of super-volcanoes – in 1971. It was 10,000 times greater than that of Mount St. Helens in 1980. Scientists disagree about its impact.

HOW VOLCANOES WORK

Under the surface of the Earth lies a thick layer of very hot, semi-liquid rock, called magma. The continental landmasses and the ocean floors are on 'plates' of rock which sit on top of the magma. At or near the point where the plates join, and at occasional 'hotspots', magma leaks through or is forced through to the surface, where it is known as lava. Some volcanoes produce a steady stream of lava; others build up a huge supply of it in an underground chamber. The pressure exerted by the build-up of magma may become so great that the chamber explodes in a violent eruption.

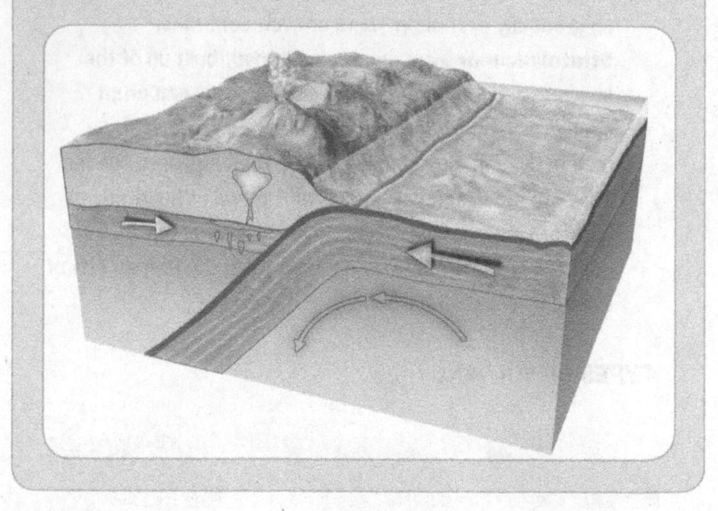

Some of them link a 1,000-year global cooling period with the eruption, and suggest it was responsible for a bottleneck in human evolution around 70,000 years ago when the population was reduced to only 1,000–10,000 breeding couples. Other animals also show evidence of a genetic bottleneck at that time.

Perhaps humans have already narrowly escaped extinction-by-super-volcano once, driven to the brink by the ensuing

MEASURING ERUPTIONS

The intensity of volcanic eruptions is measured on the Volcanic Explosivity Index (VEI) from 1 to 8. This rates eruptions on the height of the column of ash, lava and smoke produced from the top of the volcano, the mass of lava and ash thrown out, and how long the eruption lasts. The eruption of Mount St. Helens, USA, in 1980 scored 5 (but only just). The eruption of Vesuvius that destroyed Pompeii was probably a solid 5. Krakatau in Indonesia, which could be heard hundreds of kilometres away and that spread dust around the world in 1883, scored 6. Tambora, which caused worldwide cooling for two years when it erupted in 1815 was a 7. Super-volcanoes score 8.

The scale is logarithmic, which means that each increase of 1 on the scale represents a ten-fold increase in the size of the eruption: a 5 is ten times as powerful as a 4, and an 8 is ten times as powerful as a 7, a hundred times more powerful than a 6, and a thousand times more powerful than a 5, and so on.

climate change. The Toba eruption threw out around 3,000 cubic kilometres/km³ (700 cubic miles) of ash and lava, which could have cooled the climate by as much as 15 °C for the first few years.

Volcanic catastrophes

There have been enough catastrophic eruptions in recorded history to give us a hint of what might lie in store for us if a super-volcano were to rouse from its slumbers.

The eruption of Vesuvius in AD79 which destroyed the Roman city of Pompeii was described by the 18-year-old Pliny the Younger. It seemed like the end of the world, as no one had any idea that Vesuvius was even a volcano until burning rock and choking ash rained down on the countryside and cities. Yet the Vesuvius eruption was only VEI 5. Lakagígar, a volcanic field in Iceland, erupted over a period of eight months in 1783–4. It produced 14 km³ (3.3 miles³) of lava and clouds of hydrochloric acid and sulphur dioxide. The poisonous fumes killed half the livestock in Iceland, leading to a famine that claimed 20–25 per cent of the country's population. People and livestock around Europe suffered from the acidic aerosol, and the weather changed dramatically. The winter of 1783–4 was the coldest for 250 years. The effects were felt as far away as Africa, India and North America. Lakagígar was the second largest eruption in the last 1,000 years; and has been retrospectively graded as VEI 6.

The volcanic island of Krakatau in Indonesia blew apart or collapsed in 1883, throwing ash 80 km (50 miles) into the air. The subsequent tsunami, up to 46 m (150ft) tall, went around the Earth three times. The ash in the atmosphere caused global temperatures to fall by 1.2 °C and produced magnificent sunsets and slightly darkened daytime skies that persisted for years. Again, it was VEI 6.

Perhaps the most violent eruption in recorded history occurred in 1815. Estimated at VEI 7, the eruption of the Indonesian volcano Tambora in April of that year produced the 'year without a summer', with extreme weather events and low temperatures worldwide.

The sound of the eruption, like gunfire, was heard 2,600 km (1,600 miles) away, and

'A dense black cloud was coming up behind us, spreading over the earth like a flood . . . darkness fell, not the dark of a moonless or cloudy night, but as if the lamp had been put out in a closed room. 'You could hear the shrieks of women, the wailing of infants, and the shouting of men. . . . Many besought the aid of the gods, but still more imagined there were no gods left, and that the universe was plunged into eternal darkness for evermore.'

Pliny the Younger, AD79

ash fell 1,300 km (800 miles) from the volcano. It was pitch dark for two days, even at a distance of 600 km (370 miles) from the site. Modern experts suggest that 70,000–100,000 people might have died, many from starvation and disease in the aftermath of the eruption. A super-volcanic eruption of VEI 8 would be ten times as powerful as Tambora.

Dead or sleeping?

It's difficult to tell whether a volcano is extinct (never going to erupt again) or just dormant (currently inactive). Some sources consider a volcano that has not erupted in 10,000 years to be extinct, but this is not entirely reliable, as some volcanoes – particularly super-volcanoes – rest for much longer than that. The Fourpeaked Mountain in Alaska had long been thought extinct, having last erupted around 8000BC – but it erupted again in 2006.

> ***'This past week, and the two prior to it, more poison fell from the sky than words can describe: ash, volcanic hairs [hair-thin strands of volcanic mineral glass], rain full of sulphur and saltpetre, all of it mixed with sand. The snouts, nostrils and feet of livestock grazing or walking on the grass turned bright yellow and raw. All water went tepid and light blue in colour and gravel slides turned grey. All the earth's plants burned, withered and turned grey, one after another, as the fire increased and neared the settlements.'***
>
> Jón Steingrímsson, priest in Vestur-Skaftafellssýsla, Iceland, 1783

Known super-volcanoes

The best-known super-volcano lies under Yellowstone Park in the USA. There are another two in the USA alone. Some super-volcanoes are thought to be extinct – such as the one under Edinburgh in Scotland.

It's not easy to track all the super-eruptions that happened long in the past; the only known VEI 8 eruptions have been in the USA (two locations), Indonesia (Toba), Chile, New Zealand

The city of Edinburgh is built on top of a volcano that last erupted 200 million years ago.

and Argentina. The largest eruption, at the San Juan volcanic field in Colorado, produced 5,000 km³ (1,200 miles³) of ash and lava – but that was nearly 28 million years ago. Experts believe there are about 20 super-volcanoes around the world.

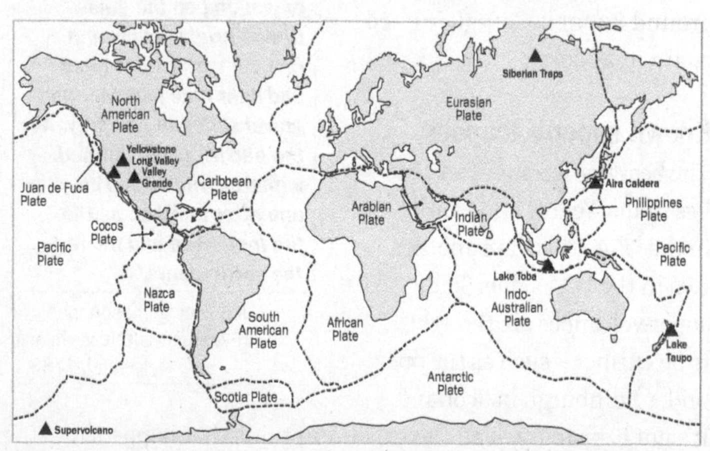

Most super-volcanoes are near the boundaries between the Earth's tectonic plates (see pages 165–6).

Prime suspect

Yellowstone, for most people, remains the prime candidate for a devastating super-eruption. It lies under Yellowstone National Park which covers parts of Wyoming, Montana and Idaho in the USA. Its latest caldera is 80km (50 miles) across. It's a beautiful area, with acres of protected forest and an unearthly landscape of brightly-coloured, mineral-enriched pools, spectacular geysers and bubbling hot mud. Yet what lies beneath is terrifying.

Yellowstone was first identified as volcanic in the 1870s but was considered extinct. In fact, it just has a very long interval between eruptions. The full, gigantic extent of the main Yellowstone caldera was mapped in the 1960s and 1970s and confirmed by satellite photography. The US Geological Survey (USGS) monitors its activity constantly.

So far, there have only been three massive eruptions in Yellowstone Park. Yellowstone is over a volcanic hotspot, which is an area in the Earth where a plume of magma rises to the surface. As the rocky plates carrying the Earth's crust move constantly, the land that is over the hotspot changes. There has been a string of eruptions from the hotspot now under Yellowstone, but apart from the last three, they were not actually in Yellowstone. There have been at least 12 super-eruptions over the last 18 million years, forming a chain of calderas across Idaho, Oregon and Nevada. The Snake River Plain was formed from the lava flows of these eruptions. The earliest known eruption from the same hotspot was 70 million years ago in Yukon, Canada.

Are we in danger?

The intervals between known previous super-eruptions range from around 4,000,000 to 300,000 years. That the last super-eruption was 640,000 years ago, after an interval of 600,000

MAXIMUM IMPACT

Although the super-eruption in San Juan, Colorado, was massive, it's dwarfed by the energy of the Chicxulub impact, the asteroid or comet thought to have helped wipe out the non-avian dinosaurs – that was 400 times more powerful.

years, gives catastrophists something to panic about. The volcano is certainly still rumbling away. The ground above the hotspot is rising steadily, a few centimetres a year, and there are sporadic swarms of small earthquakes. Recently, USGS announced that the magma chamber beneath Yellowstone is 2.5 times larger than previously thought. Minor volcanic activity is evident all the time, with geysers and bubbling hot streams that are heated by magma below the surface. Yet this is no guarantee that an eruption is imminent. Furthermore, the hotspot produces many smaller eruptions between its big blow-outs. The most recent of those was 70,000 years ago.

Currently, USGS says there is no sign of imminent eruption. They expect warning signs weeks, months or possibly even years in advance. But if we knew it was coming, what could we do?

What if . . . ?

The most recent super-eruption at Yellowstone was 1,000 times more violent than the eruption of Mount St. Helens in 1980, which Americans consider a large eruption. In a similar-sized eruption, a column of ash would rise 30 km (18 miles) into the air and deposit debris as far away as the Gulf of Mexico. Searing winds carrying fog, ash and rock superheated to 800–1,000 °C, which instantly burn all in their path, would ravage the land. And they travel at 700 kmh (435mph), so there's no running away from them. Ash and lava would fill nearby valleys to a depth of more than 100 m (300 ft), the debris so hot it would weld itself together, filling in the valleys with solid rock. Some estimates say 100 million North Americans would die in the weeks immediately following a super-eruption.

Gases and ash rising high into the atmosphere would mix with water vapour producing a haze to block out sunlight

IT'S ALIVE!

The discovery that Yellowstone is still active came in 1973. Geologist Bob Smith had worked on Peale Island in Yellowstone Lake in 1956. When he returned in 1973, intending to use the same dock for his boat, he found the dock under water; nearby trees along the shoreline were dying and partly submerged. Checking benchmarks that had been placed on roads since [illegible]3, he found that the area, at the north end of the lake, had [illegible] 75 cm (30 in) over that time. The south end had not risen [illegible] the ground was doming over the volcano. It was, in [illegible]ords, a 'living, breathing caldera'.

for years or decades, plunging Earth into a volcanic winter that would kill plants and animals alike. Farming would be unproductive everywhere in the world. If suggestions that the eruption of Toba caused a 1,000-year-long cold snap are correct, the next ten centuries could be blighted.

EXTERMINATION

The Deccan Traps form a vast plateau in India. They are a slab of rock 2,000 m (6,500 ft) thick, with a total volume of 512,000 km³ (123,000 miles³) made from solidified lava that erupted 66 million years ago. The eruptions might have lasted for 30,000 years, and covered an area half the size of India with lava. Global temperatures fell 2 °C, and many experts think that the combination of the eruptions and the Chicxulub asteroid sealed the fate of the dinosaurs, exterminating 75 per cent of plant and animal species on Earth.

CHAPTER 9

Could we live for a thousand years?

People are living longer – but extending life by a few years is nothing compared with the ambition of extending it by nine centuries.

Threescore years and ten

The 'normal' human lifespan is often estimated at around 70 years. But it is frequently much shorter than that, especially when people are threatened by dangers such as war, famine and disease. Neverthless, our potential to live into old age has always been clear from the extended lives of the privileged and well-fed. Among the Ancient Greeks, Socrates was 70 or 71 when he was executed, and Isocrates lived to 98; in Ancient Egypt, more than 3,000 years ago, Ramesses II lived to 90.

'Normal' lifespan, of course, is a vague term. It's not the maximum possible, nor the average, but perhaps what we would expect if someone was lucky in terms of avoiding disasters and had a reasonable quality of life. In the past, infant and child mortality were much higher, and many more people died in their teens, twenties, thirties, forties and fifties than now, so pushing down the average lifespan. Because of the dangers of childbirth, women were most at risk between puberty and their forties, . This was also a risky age for men, who were subject to disease, accidents, and death in battle.

Now, mainly thanks to childhood vaccination, safer childbirth and work practices and fewer wars, most people in industrialized nations expect to live until their 70s or 80s at least. A fair number will make it to 100 and a few to 110 or even 115. But what if we could live much, much longer? Say, to 800, 900 or even 1,000. Could it ever be possible? What would be the social implications? And would we even want to live so long?

'The days of our years are threescore years and ten; and if by reason of strength they be fourscore years, yet is their strength labour and sorrow; for it is soon cut off, and we fly away.'

Psalm 90, King James Bible

Why die?

Leaving aside mythological figures said to have achieved great age, only a tiny proportion of people have ever lived much beyond 100 years. There seems to be something innately self-limiting about the human body. There have been several studies to try to discover factors that affect or cause ageing. Some that have shown clear-cut results investigated food intake; others looked at the structure of chromosomes.

Eat less, live longer?

Studies in rats and mice have found that restricting calorie intake while at the same time guarding against malnutrition can extend lifespan – sometimes doubling it. One study in 2014 found the same advantage of a restricted diet in primates. Twice as many rhesus macaques survived to 35 years of age on a calorie-restricted diet as on a normal diet (the control group). (Deaths from causes not related to age were excluded from the results.) This suggests that we could live longer if we ate less – and not just sufficient to avoid being overweight or obese, but less than we might habitually eat to stay a healthy size.

Experiments into calorie-restricted diets are carried out under controlled conditions, and generally the animals are given dietary supplements to make sure it is only calories and not other essential nutrients that are restricted. The food supply for the macaques was reduced gradually by 30 per cent over a period of three months, allowing them to adapt to the change. Calorie-restriction experiments have also been carried out with human subjects, but with mixed results.

How fit are your cells?

Another approach to the way cells deteriorate through ageing (senescence in biology-speak) looks at the process of cell division and the limits on it. Our bodies constantly renew themselves by growing new cells to replace those that are worn out or damaged. To produce new cells, existing cells split in half. This process is called mitosis.

WHERE TO LIVE THE LONGEST

Shimane Prefecture in Japan has more people aged over 100 per million of the population than anywhere else in the world. In 2010, there were 743 centenarians per million people. Japan, the USA and France have had some of the longest-living individuals. The oldest person whose age could be conclusively proven was Jeanne Calment, who died in France at the age of 122 years (1875–1997).

During mitosis, the cell duplicates its contents, then each set of cell components separates and a cell wall grows between them, dividing the two sides into two separate cells.

The chromosomes are long strands of DNA which encode the genetic instructions for making the organism (see *What's the difference between a person and a lettuce?* page 61). DNA is a very long and complex molecule that looks rather like a ladder twisted into a spiral (see *Could we bring dinosaurs back to life?* page 61). All the way along, the 'rungs' are made up of bases. Preparing to duplicate, the DNA 'unzips' into two halves. Each half is easily rebuilt as each base just has to be matched with its usual companion. The cell builds the second half of each strand and then has two sets of identical chromosomes.

So far, so good. But the mechanism by which the cell does this rather complex task involves dropping a bit from the end of each chromosome. This means that each chromosome reduces in length every time it is duplicated. That could be disastrous, but luckily the chromosomes come prepared. Each has a 'cap' of non-coding material at the ends called a telomere. This protects the important coding part of the chromosome so that all the information can be passed on intact. However, each time the cell divides and the chromosomes are duplicated, part of the telomere is lopped off.

Eventually, after many duplications, there is too little of the telomere cap left to protect the chromosome. At this point the cell becomes senescent or dies. If the cell were to carry on

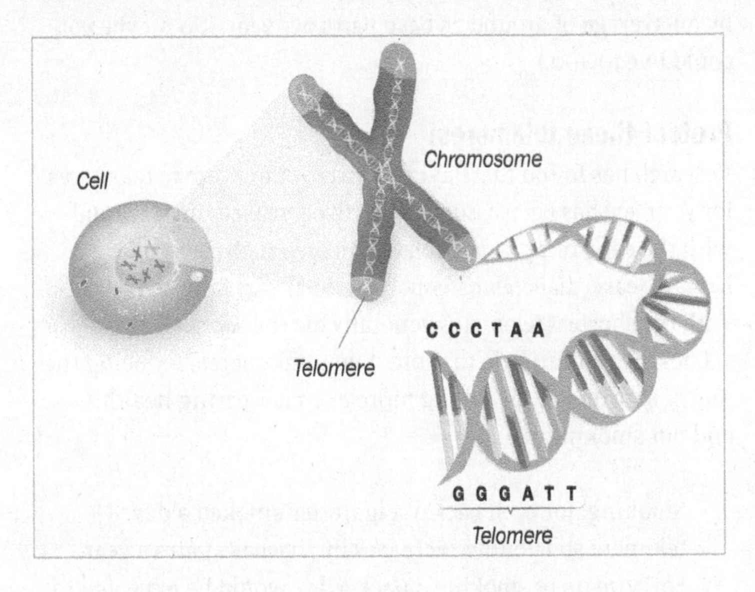

duplicating, it would lead to damaged DNA that would result in malfunctioning cells, which would be dangerous for the whole organism. If this happens and faulty cells carry on duplicating, it can lead to cancer.

So, as an organism ages its telomeres shorten. That's fairly obvious – if they lose a bit each time the cells divide, the older the organism is, the more shortenings will have occurred. It's possible to measure the rate of ageing by looking at how much the telomeres have shortened, as long as we know the original optimal telomere length for that organism. For young humans, it's around 10,000 nucleotides. A nucleotide is a group of base pairs; in telomeres, nucleotides always have the configuration TTAAAGG (that is thymine, thymine, adenine, adenine, adenine, guanine, guanine). The rate of shortening varies; human liver cells lose around 55 base pairs per year. (This means that if you could live until your liver cells ran out of telomere you might make it to about 180.) Other cells shorten by an average of around 25 base pairs per year. (So maybe you could live to 400.)

Protect those telomeres!

Research has found that having shorter-than-average telomeres for your age has been associated with decreased lifespan, and with the early onset of age-related disease such as coronary heart disease, diabetes, osteoporosis and heart failure. People with the shortest telomeres generally die the soonest.

Luckily, you can help to protect your telomeres, reducing the rate of shortening, by taking more exercise, eating healthily and not smoking.

- Smoking: for each pack of cigarettes smoked a day, telomere shortening increases by five base pairs a year. Forty years of smoking a pack a day would be expected to reduce lifespan by 7.4 years (on average).
- Obesity: telomeres are considerably shorter in obese individuals than in slender people of the same age. The age-cost of obesity as a result of telomere shortening has been calculated at 8.8 years.
- Pollution: excessive exposure to pollutants is associated with telomere shortening. Traffic police have been found

to have shorter telomeres than office workers of the same age.

- Stress: consistent stress can lead to telomere shortening equivalent to ten years of life.
- Exercise: being active reduces telomere shortening – at least if you're very active.
- Diet: the best diet for retaining healthy, long telomeres is – guess what – rich in fibre, low in polysaturated fats, containsomega-3 fatty acids and antioxidants such as Vitamin C, Vitamin E and beta-carotene. Restricted protein intake is associated with reduced growth rate, reduced adult appetite, and reduced telomere shortening. It extends lifespan by up to 66 per cent in laboratory animals. If it works in humans, that could push our lifespan to 150 years. Protein is an essential part of our diet, though, and individual requirements vary. Don't restrict your protein intake or that of your children without medical advice.

Or not . . .

A study in 2014 found no link between body mass, smoking, exercise rates or alcohol intake and telomere shortening. The jury is still out on this one. But no one has suggested that eating healthily and being active are actually bad for you and your telomeres. A study in Tel Aviv found that caffeine increases telomere shortening but alcohol reduces it, so one offsets the other. But their study was with yeast, so might not hold for us.

The enzyme telomerase prompts telomeres to regenerate. Normally, telomerase is present in stem cells (see *Are stem cells the future of medicine?* page 141), including those in the bone marrow that produce blood cells. Stem cells are the basic cells from which all other cells in the body are derived. Telomerase is not active in body cells.

If telomerase is applied to cells in which the telomeres have been shortened over time, the telomeres regrow. This has potential for treating or delaying the onset of age-related conditions. The first clinical trials are likely to be in treatments for people with conditions that cause accelerated early ageing rather than in people who would like to live to 200 or who have been reckless with their telomeres through years of over-eating and smoking.

An experiment at Stanford University, California, in 2015 involved prompting cells to make telomerase. Researchers found they could add back 1,000 nucleotides onto shortened telomeres – that's the equivalent of many extra years of human life, perhaps a further 10 per cent life expectancy. Human skin cells treated in this way could divide around 40 more times than untreated cells.

Guru of longevity: Aubrey de Grey

Aubrey de Grey is a biomedical gerontologist who believes that extending human life considerably is scientifically possible. He has said that the first person who will live to be 1,000 has quite possibly already been born. De Grey maintains that cumulative damage to mitochondrial DNA makes a significant contribution to ageing. He has listed seven types of physical deterioration at the cellular or microscopic level which he believes contribute to ageing. He has also founded a non-profit organization to try to find treatments to tackle them, with the intention of extending human lifespan considerably.

Few scientists believe that de Grey's list really defines ageing, and it does not currently offer any kind of treatment options that would extend life. The conclusion reached by most scientists working in the area has been that there is little to support his theory, but it has not been proven to be conclusively wrong.

TELOMERES AND CANCER

Cancer cells typically have short telomeres initially, but many are then acted on by telomerase. As this extends the telomeres, the cancer cells are given a new lease of life. The cancer cells are then able to multiply rapidly without further genetic change. That makes them more successful than somatic (body) cells – and this is bad news. However, if telomerase is implicated in the unhampered growth of cancer cells, this suggests that a treatment which aimed to inhibit the function of telomerase – at least in the area of the cancer – might reduce the rate of spread or the growth of the cancer.

As with all scientific developments, it's easy to get carried away by the exciting prospects of breakthroughs and new knowledge. But extending lifespan is only worthwhile if the extra years will be healthy and active. That is the prospect held out by harnessing telomerase. Inevitably, any such treatment would be expensive, so its uptake would be limited to wealthy people in wealthy nations, at least to start with. The social impact would be significant, and possibly not in a good way.

And what about the impact on the very elderly individuals themselves? Would you really want to live centuries after your loved ones had died? Would you want to see the world you had known disappear completely?

Live longer or come back from the dead

With no immediate prospect of hugely extended life or immortality, some people are paying vast sums of money to have their bodies preserved cryogenically (that is, frozen after death). Those with a little less money have opted just to have their head preserved. The idea is that at some point in the future, when whatever they have died of can be successfully treated, these people will be revived, cured, and can carry on living. So far, over 150 people in the USA have been cryogenically preserved (300 worldwide) and a further 80 heads are in storage.

But again – how would it feel to be revived 100 years after everyone you have ever known has gone?

CRYOGENICS

There is no DIY option for cryogenic preservation – you can't do it with a home freezer.

As soon as the person has died, the cryogenic team tries to keep the blood circulating while they move the body to the preservation plant. There the body temperature is reduced until it is near 0 °C. The blood is drained from the body and replaced with a cryoprotectant solution to prevent ice crystals forming in the organs and tissues and damaging them. The corpse is then cooled to -130 °C and put into a container, which is lowered into a tank of liquid nitrogen. It is kept at a steady temperature of -196 °C. The cooling process is done very slowly, taking two or three weeks at around half a degree per day.

Critics point out that the conditions and cryoprotectants needed vary for different organs of the body, that we currently have no technology for safely defrosting frozen people, and that anyone successfully defrosted might suffer serious physical damage and, possibly, complete memory loss. A body at -196 °C would be so brittle that it could shatter like glass if shocked, and even the process of defrosting might produce sufficient thermal shock to break it irrevocably. If you're still keen on the idea, it costs around $200,000 for a full-body preservation, or $80,000 for the head alone.

CHAPTER 10

Why don't satellites fall down?

Satellites are within range of Earth's gravity, yet as a rule they don't fall from the sky – why not?

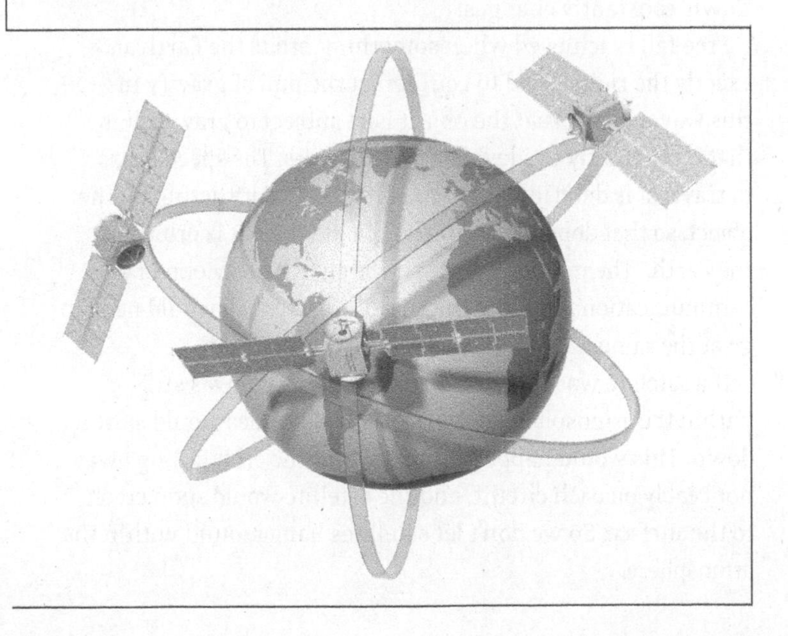

Floating

Earth's gravity extends well beyond the atmosphere – after all, the Moon is held in orbit and doesn't escape into space. The International Space Station (ISS) is a satellite that orbits the Earth just as weather and communications satellites do. The weightlessness that astronauts experience there is a clue as to why the satellites stay in orbit.

Falling without landing

Satellites don't crash back to Earth for the same reason that astronauts are weightless on the ISS. Both are in free-fall.

Free-fall is a state of perpetually falling, but never actually hitting anything because 'down' has moved by the time you get there. Imagine fragmenting the act of falling into tiny interludes of time. Each moment the object falls towards the centre of a massive body (such as the Earth). But each moment it is also moving around the body, so the place it is aiming at as 'down' constantly changes.

Free-fall is achieved when something orbits the Earth at exactly the right speed to counteract the pull of gravity in this way – it isn't that the object isn't subject to gravity, but that it constantly evades the consequences. The speed it has to travel at is determined by the force of gravity acting on the object, so that depends on the height at which it is orbiting the Earth. The mass makes no difference – the Moon or a communications satellite the size of a basketball would need to go at the same speed if it was at the same altitude.

If a satellite was in too low an orbit, so that it was still within the atmosphere, drag from air molecules would slow it down. This would happen quite quickly, the orbit falling away noticeably on each circuit, and the satellite would soon crash to the surface. So we don't let satellites hang around within the atmosphere.

WEIGHT AND MASS

Weight and mass are not the same thing, though we often use the words interchangeably. Weight is the effect of gravity acting on mass. That means an object of a fixed mass, say 10kg (22lbs), will be heavier on Earth than somewhere with lower gravity, such as the Moon.

On Earth, your weight and your mass are the same because our reference point is the same – Earth's gravity. Astronauts on the Moon have the same mass as on Earth, but their weight is around a tenth of what it is here. Consequently, an astronaut on the Moon can lift more massive objects (that is, those with greater mass) than on Earth, because the objects weigh less.

Pick your orbit

Orbits are divided into bands, with Low Earth orbit defined as anything below an altitude of 2,000 km (1,200 miles); Medium Earth orbit at 2,000–35,786 km (1,200–22,236 miles); and High Earth orbit above 35,786 km. Most satellites are in Low Earth orbit, at an altitude of around 650 km (400 miles).

The precise altitude of 35,786 km is a special place, called a 'geosynchronous orbit'. Here, the satellite is travelling at a speed that exactly matches the rotation of the Earth. The effect is that the satellite keeps up with a point on Earth, remaining above it at all times, where it is said to be 'geostationary'. Communication satellites are usually geostationary, enabling them to relay signals between fixed points on Earth.

Beyond the geosynchronous orbit is an area known as the 'graveyard orbit', where old satellites are sent to die. Here, they just carry on orbiting, out of the way, endlessly. It's getting quite crowded up there. Space junk – bits of satellite, whole satellites, bits dropped from spacecraft – is becoming an increasing danger to spacecraft launched from Earth, which have to make their way through the graveyard.

Satellites such as the ISS, not in geosynchronous orbit, move over the surface of the Earth. The ISS travels at around 27,600 kph (17,000 mph) at an altitude of 330–435 km (205–270 miles). One effect of this is that the crew sees a sunrise every 90 minutes! Non-geostationary satellites are used for applications such as surveying and weather forecasting.

More fun with gravity

Isaac Newton's law of universal gravitation, published in 1687, established that the force of gravity is inversely related to the square of the distance between two objects.

In simple terms, this means that if you double the distance between two objects, the force of gravity operating between them is a quarter of what it was. When measuring the Earth's gravitational pull, the centre of the Earth is the starting point for the measurement, though of course the gravity we experience every day is that at the Earth's surface, 6,371 km (3,958 miles) from the centre.

Cancelling gravity

As the force of gravity reduces the further we get from a body such as the Earth or Moon, there is a point between the two where the Moon's gravity and the Earth's gravity are exactly equal and balance each other out. An object released at this point with no acceleration would remain suspended between the two. This neutral point is also known as a Lagrangian point, named after the Italian astronomer Joseph-Louis Lagrange (1736–1813), who first calculated it.

There is also a Lagrangian point between the Earth and the Sun. Satellites have been placed at Lagrangian points to study the Sun and to map the universe. The Moon's gravity becomes

WHICH WAY IS DOWN?

'Down' is always towards the centre of the Earth. Whether you drop an item at the equator, or in Australia, or at one of the poles, it will always fall to the ground because it is attracted towards the centre of the Earth. In space, objects are attracted towards the centre of any bodies that are exerting a gravitational force. This means there can be lots of sources of gravity pulling an object in different directions. The greatest gravitational pull might not be from the nearest object, as it is related to mass as well as distance. A spacecraft released between the Earth and the Sun might be closer to the Earth, but still drawn into orbit around the Sun, if this is exerting the greater gravitational force. Although Earth is always shown in space with north at the top, so it looks as though south is 'down', this is simply convention: up, down, north and south have no meaning in space.

The Sun rising above the curve of the Earth, seen from the International Space Station.

stronger than Earth's, the closer a satellite gets to it. An object released at this point, with no acceleration, would fall towards the Moon rather than the Earth and would eventually drop into orbit around the Moon (or crash into it).

Earth's mass is around 81 times that of the Moon. As gravity follows Newton's inverse square law, the neutral point between Earth and the Moon is about nine times as far from Earth as from the Moon (because $9^2 = 81$). This is around 340,000 km (210,000 miles) from Earth. It's not a single, stable point, though. The equations which allow us to calculate its location assume that each body is a perfectly smooth sphere, but neither the Earth nor the Moon is properly spherical or smooth. There is also variation in the density of each, which means gravity is not equal all over the surface. In addition, the orbit of the Moon is not truly circular, but rather elliptical. And finally, the exact location of the neutral point is affected by which part of the Earth is closest to the Moon at any time. If the Himalayas are between the Earth and the Moon, for example, this will affect where the neutral point lies.

As the Moon is orbiting the Earth, and both are moving through space around the Sun, the neutral point also orbits the Earth, lying on a line between the centre of the Earth and the centre of the Moon at any particular moment.

More stable places

There is more than one neutral point relating to two bodies, if one is orbiting the other. There are five Lagrangian points, of which the most obvious – the one just described – is known as L1. The others are:

- L2, which is outside the orbit of the smaller body, the same distance from it as L1 but opposite it
- L3, which is on the far side of the larger body, on the

The 'Potsdam potato' is an image of the Earth drawn to map the strength of gravity at different points.

orbitable path, directly opposite the current position of the smaller body
- L4 and L5, which are at the points of equilateral triangles drawn with a line between the two centres as their base.

Of these, L4 and L5 are the most stable. Objects orbiting at L1, L2 and L3 tend to fall out of orbit.

Let's live in space

At points L4 and L5 in particular, an object can follow the orbit of the smaller body using no power of its own. This makes them excellent places to park space stations. A number of observatories and space telescopes occupy the L1 and L2 points for the Sun–Earth system.

Several planets have natural satellites in the L4 and L5 positions relative to the Sun. (As the L1–L3 orbits tend to be unstable, natural satellites in those positions are not common. The instability comes from the gravitational pull of other planets in orbit around the Sun.) These natural satellites are generally asteroids, though Saturn's moon Tethys has two smaller moons at its L4 and L5 points, called Calypso and

Telesto. Jupiter has two hosts of asteroids, one at each point, known as the Greek and Trojan camps.

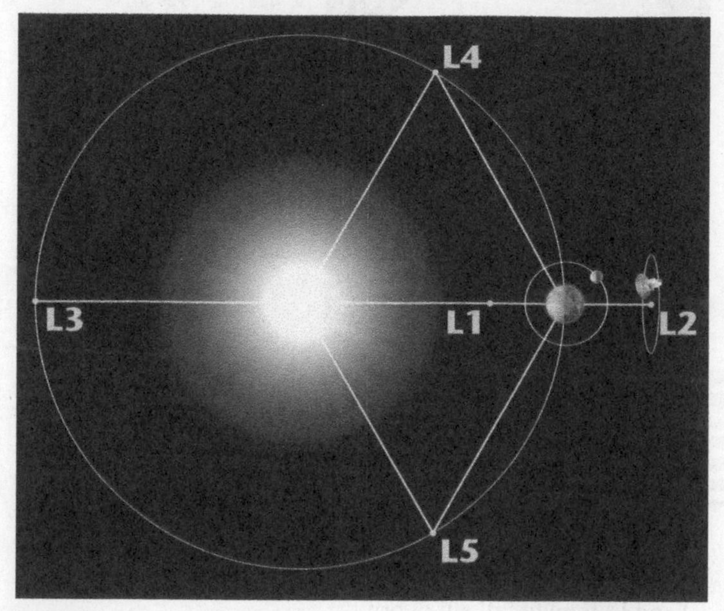

Lagrangian points for the Earth and the Sun.

Going up...

Most of the energy (and so fuel) used to launch a satellite is not actually needed to send it upwards, but to accelerate it to the right speed for its orbit. Space is not really very far away – it starts at 100 km (62 miles) above the ground. That means that some places – such as Seattle, Beijing, Cairo and Canberra are closer to space than to the sea. The real challenge is to accelerate a vehicle to a speed of around 8 km (5 miles) per second to keep it in space. Once in orbit, it doesn't need to burn fuel all the time. It just needs a little energy now and then to right it if it starts to drift (called 'station keeping').

... and down

Sometimes the orbit of a satellite will decay. This generally happens because it loses speed, so is no longer able to keep ahead of gravity's pull. Most falling satellites burn or break up as they enter Earth's atmosphere. The friction between air molecules and the material of the satellite produces sufficient energy to destroy the satellite. Just occasionally, large chunks fall to Earth intact. This is why satellites are often 'parked' in the graveyard orbit – so they don't crash to Earth and cause any damage.

What about relativity?

Newton's model of gravity as a force has been superseded by Einstein's explanation in the theory of general relativity. Einstein explains gravity as a feature of curved space-time. This is most commonly depicted as showing space-time as a blanket held taut, with massive bodies (balls on a blanket, or planets and stars in space-time) causing a dip, or curvature. Other objects naturally fall towards the heavier bodies, being drawn into the dip. It's not a perfect analogy, as it shows four-dimensional space-time as a two-dimensional blanket in three-dimensional space, but it's a good start.

A diagram that shows the curvature in three dimensions can be harder to understand but is closer to the truth (see right).

Luckily, the behaviour of satellites follows Newton's laws and the discrepancy between Newton's and Einstein's accounts doesn't make much difference at this scale. Newton's laws falter, though, when trying to deal with very tiny or very large systems.

THE L5 SOCIETY

Formed in 1975, the L5 Society proposed to build giant space colonies at the L4 and L5 points of the Moon's orbit around the Earth. Needless to say, these have not so far materialized. The L5 Society merged with the National Space Institute in 1986, existing today as the National Space Society.

CHAPTER 11

What would happen if you fell into a black hole?

You're unlikely to fall into a black hole on your way to work. But what would happen if you did?

Neither black nor a hole

'Black hole' is a bit of a misnomer. Generally, a hole means there is an absence of something, yet in a black hole there is a very great deal of something. The official definition of a black hole is a region of space-time which exhibits extremely strong gravitational effects. As gravity is exerted by matter, acting on other matter, the power of gravity a body exerts is in proportion to its mass. This suggests, correctly, that since a black hole produces extreme gravity, it has enormous mass (is very massive, in the technical sense of the word).

The volume occupied by a black hole is very small for the amount of mass it has, so the black hole is very dense. Black holes form when matter collapses in on itself, producing something so dense that its gravitational pull is too strong to allow even light to escape it. This means that black holes can appear as black areas of space – they can look like 'holes' in the canopy of stars.

How we 'see' black holes

Because black holes emit and reflect no light, they are invisible. This makes them difficult to spot. Astronomers can detect their presence, though, from their impact on other objects. A black hole near a star sometimes affects the orbit of the star. Light from a star or galaxy behind the black hole can be bent by the black hole's gravity, an effect called gravitational lensing. This means the star or galaxy can be seen by astronomers, when it should in fact be hidden. And if a black hole has a strong enough gravitational pull to draw in gas from a nearby star or other matter, this heats up as it spirals into the black hole. It then emits a bright burst of radiation which space telescopes can pick up. Not all black holes will be displaying these tell-tale signs right now, so there are probably a lot of black holes out there that we don't know about.

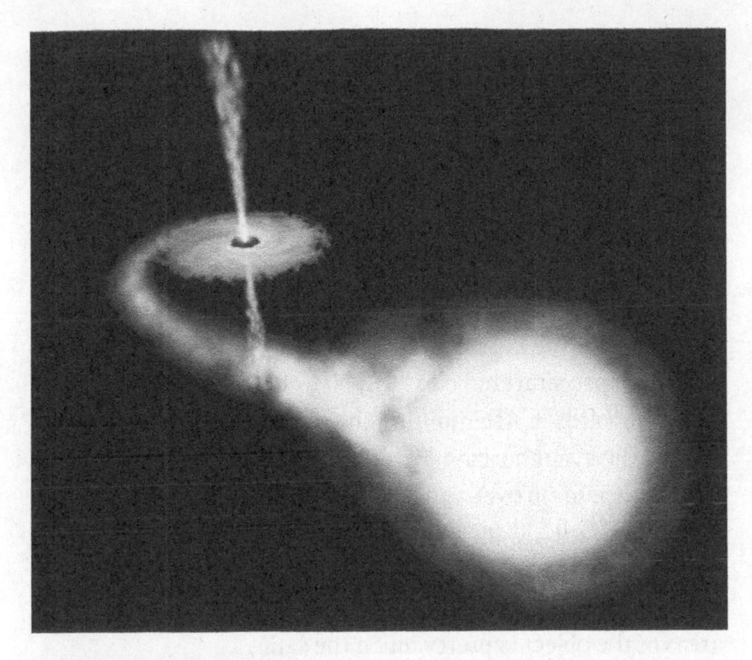

Supercharged gravity

It is the extreme gravity of black holes that makes them what they are. Around each black hole is a boundary called the event horizon. Once anything crosses the event horizon, the pull of the black hole's gravity becomes irresistible and the object (or light) is pulled in towards it, inevitably becoming part of the black hole.

Not everything drawn towards the event horizon of a black hole actually ends up as part of the black hole, however. Matter in orbit around a black hole, outside the event horizon, forms an accretion disc. Some of this, as it loses its angular momentum, will fall towards the black hole. Matter spinning round in the accretion disc bumps into other particles at very high speed, and some matter and radiation are blasted back out into space. This is called an outflow, and can provide another clue that the black hole is there.

Falling in?

The gravity of a black hole is not infinite. Gravity is a function of mass, and the gravitational pull exerted on another body depends on the mass of them both and the distance between them (see *Why don't satellites fall down?* page 91). Just as the Earth doesn't fall into the Sun, a body straying near or orbiting a black hole will not inevitably fall into the black hole.

However, just as a body that strays too close to the surface of the Sun (or the Earth) is drawn in by its gravity, if we were to send a spacecraft too close to a black hole, it too would be drawn towards it. After joining the accretion disc whirling around the event horizon, it could lose sufficient angular momentum to tip over and be pulled towards the centre of the black hole. As it did so, a strange thing would happen.

If we watch something fall to Earth under the influence of gravity, the object keeps its shape. The impact of gravity on all areas of the object is pretty much the same.

In the case of a black hole, however, the gravity is so immense that even over a short distance, such as the length of a spacecraft or the height of an astronaut, the changing effects of gravity with distance have an impact. The part of the craft (or astronaut) closest to the black hole – the bit going in first – would be subject to greater gravitational pull than the part furthest from it. This means that one end would accelerate towards the black hole more quickly than the other end, resulting in stretching or 'spaghettification': the object being drawn out into a very long thin strip. This would only happen for a moment – then the object would disappear into the void and become part of the black hole's super-dense matter.

If you stay on Earth and don't go wandering space in the vicinity of black holes, you're safe from being sucked into one. Black holes are in fixed places (or orbits). When a large star goes supernova and leaves behind a black hole (see box on page 105),

HOW TO MAKE A BLACK HOLE

There are thought to be three types of black hole, distinguished by size. The smallest are primordial black holes, formed soon after the Big Bang from which the universe formed. They can be as tiny as a single atom, with a mass of many millions of kilograms.

Stellar black holes are of medium size; they form when large stars die. The star typically explodes in a spectacular supernova, throwing out light and matter. The rest of the star collapses in on itself, leaving a black hole where the star once was. The black hole has much of the mass of the original star, but occupies a tiny space. A black hole with a mass 20 times that of our Sun might be only 15 km (9 miles) across.

The largest are supermassive black holes. Like primordial black holes, they are likely to be very old, probably forming at the same time as the galaxies they are situated in. Most, if not all, large galaxies are thought to have a supermassive black hole at their centres. The black hole at the centre of our Milky Way is known as Sagittarius A*; it is about the size of the sun, but has four million times its mass.

the black hole occupies the centre of the space previously occupied by the star. It shares the same field of gravitational impact that the star had, so will not suck in nearby stars or planets – anything vulnerable would already have been drawn in by the star's gravity.

Astronomers believe that the only new black holes that will form now will come about from the collapse of stars. It's too late for primordial black holes to form, and supermassive black holes only do so with new galaxies. Our own Sun is too small to become a black hole. It would require a star twice the size of our Sun to take the supernova/black hole route to death.

CAN WE MAKE A BLACK HOLE?

Scientists at CERN, a research facility on the border between France and Switzerland border, have built a huge particle accelerator called the Large Hadron Collider in order to investigate sub-atomic particles. There has been much speculation in the non-scientific press that operations at CERN might create a microscopic black hole that would gobble up the Earth or even the universe. There is absolutely no danger of this happening, for three good reasons:

- **If such microscopic black holes can be created by colliding particles, there will be many millions of them created naturally and none has harmed the Earth yet.**
- **If we could create such a microscopic black hole, it would be so unstable that it would decay in such a minuscule interval of time – one that is literally meaningless in terms of the laws of physics – that it would have no chance to do any damage.**
- **If all the laws of physics are wrong, and we could both create such a black hole and sustain it for a reasonable amount of time, it would grow so slowly, 'eating' one sub-atomic particle at a time, that it would take trillions of years to grow to a mass of one kilogram. That's much longer than the current age of the universe.**

CHAPTER 12

Why can't you uncook an egg?

Cooking tends to be a one-way street. Once you have baked a cake or cooked an egg, there's no going back.

Chemical and physical changes

If you melt ice cream or ice by warming it up, all is not lost - you can freeze it again and it will be pretty much back to what it was. Other changes with heat are often permanent, however - cooking is not something that can usually be reversed. The difference is what happens at the molecular level.

Melting ice - or ice cream - is a physical change; the molecules of water in the ice are not altered when you heat the ice to melting point. Change from a solid to a liquid, or a liquid to a gas (and back) is called a phase change. Heating a substance involves supplying energy to the molecules so they move around more. As ice is heated, the molecules move more vigorously and eventually break out of the crystalline structure that holds them in the solid structure of ice. As it is heated more and more, the molecules move even more until the water boils. At that point, molecules escape the surface of the water into the air and the water evaporates.

You can heat and cool ice/water any number of times and the water molecules do not change individually - they just move more or less depending on the temperature and phase (solid, liquid or gas).

Cooking an egg is a different matter. When you heat an egg, the protein in the egg denatures. This means that it undergoes chemical changes that alter the shape of the molecules.

Protein molecules are long strands that have to be folded into the right shape to do their job properly. They are held in place by various forces such as hydrogen bonds - weak bonds between hydrogen and some other atoms in the molecules. When the protein is heated, the molecules vibrate or shake, breaking the hydrogen bonds. Then the strands of the protein molecule unravel - the protein loses its essential shape. As its shape changes, the physical properties of the protein also undergo a change.

When we heat eggs, breaking the hydrogen bonds leads to the egg white becoming opaque and solid. Once the hydrogen bonds have broken, they can't be put back in place.

Taking the sugar out of tea?

Some changes might look as though they are irreversible, but they are not. When you dissolve sugar in water, you appear to gain a new substance (sweet water), but in fact all you have done is scatter the sugar molecules among the water molecules. No bonds have been broken or made: the sugar and the water are just very thoroughly mixed together. Sugar, like ice, has a crystalline structure. Both the water and the sugar molecules are polar, meaning they have an unequal electrical charge. As a result, areas of the sugar molecules are attracted to areas of the water molecules and so the sugar dissolves, losing its crystalline structure in the process. (If you drop sugar into olive oil, which does not have polar molecules, the sugar won't dissolve.) The water and sugar can be separated again easily as the molecules have not reconfigured or made new compounds. If you boil off the water, or just let it slowly evaporate over time, the sugar will be left as a solid, recrystallized in the bottom of the container.

Separating tea from water is similar – you can boil off the water (and condense it if you want to get the water back), and there will be a rather nasty brown stain from the tea – you've probably seen it in a tea-cup you've left lying around for too long. That's what came out of the tea-bag or tea-leaves, but it's rather hard to put it back in. It's not perfect – oils from the tea will have been lost or changed, and that's why instant tea made by freeze-drying doesn't taste the same as fresh tea.

Turning back the clock

As a very general rule, chemical changes cannot be reversed and physical changes can be. It's not entirely reliable as a guide for everyday behaviour, though. Smashing a favourite glass tumbler is a physical change, but you can't put it back together exactly as it was before. (In terms of science, it is reversible because you can re-melt the glass and remould the tumbler again, but effectively it is a new tumbler made with the old glass.) Some chemical changes can be reversed. Indeed, your life depends on the reversibility of chemical changes. Oxygen from the air you inhale combines with a protein in the blood called haemoglobin to form a new compound, oxyhaemoglobin. The blood carries oxyhaemoglobin around the body and delivers it to the tissues that need oxygen. At that point, the oxyhaemoglobin breaks down into haemoglobin and oxygen again, neither being the worse for wear. The haemoglobin is carried back to the lungs to pick up its next consignment of oxygen.

CHAPTER 13

Can we talk to the animals?

Some animals can mimic human speech and others understand key words. So what is the exact nature of this human/animal communication?

Talking and not talking

Some animals make noises automatically, without having to learn them; other animals learn sequences of sounds through repetition. The animals that make sounds immediately from birth or hatching are responding to genetic programming rather than learning any kind of 'language'. They tweet, growl or hiss, usually in response to emotional triggers such as fear, in the same way that a duckling swims automatically when it encounters water, without having been taught.

Humans are among the animals that learn new sounds and add them to their innate vocabulary. All babies cry to communicate their needs, but slowly they learn to use other sounds and start to make words. This use of language is special to humans, even among primates. With chimpanzees, verbal communication involves using sounds they know innately. Their combination of grunts, screeches and hoots does not develop with age. Although the voice of an adult chimp is lower than that of a juvenile, the vocalizations are the same. As far as we can tell, their range of verbal communications is quite limited.

The ability to learn a language or extend the range of sounds that we and other animals make is relatively rare. Animals which are able to do this derive from vastly differing species, which suggests that the characteristic has evolved separately on several occasions. Vocal learning is found in three groups of birds – parrots, songbirds and hummingbirds – and five groups of mammals – elephants, seals, bats, cetaceans (whales and dolphins) and humans.

Why we talk

The way in which we speak is the result of a combination of factors: suitably adapted vocal cords, the ability to control our breathing, and the type of brain we have. Our brains can learn and mimic new sounds and use language with a consistent grammar. Those animals that are good at mimicking human speech – most notably, parrots – have the same mental ability to learn new sounds.

Programmed to sing

It has long been known that zebra finches learn to sing by imitating adult male finches. If young finches are raised in isolation, unable to hear the males they model their song on, they don't learn to 'speak zebra finch' – they don't produce the normal songs used by their species. Instead, they produce an irregular, raspy sound.

If these isolated birds are then raised to adulthood and allowed to mate, they teach the same odd song to their own offspring. After a few generations, the isolated population begins to produce songs that are close to normal. By the fifth generation, it sounds very like the song produced by wild zebra finches. This suggests that a combination of learning and genetics acts to produce communication in songbirds.

Obviously, it would be unethical to try to reproduce this experiment with human children (see box on page 115). However, there are well-documented accounts of twins developing their own private language (a phenomenon called 'cryptophasia'). It's relatively rare, but when twins do develop a private language, it always has the same grammatical structure. This is a very simple structure and doesn't allow for complex or nuanced statements. It tends to refer only to immediate situations, making no distinction between subject and object and having no words to denote other locations in time or space – everything is in the here and now. It's impossible to tell whether the language would eventually become more sophisticated, because twins soon replace their private language with one they can use to communicate with other people. Deaf children raised without being taught a sign language also develop their own sign language, and this, too, follows the grammatical rules of twin-languages.

Sense in sentences

Titi monkeys are small primates that live in the tropical forests in South America. If worried or frightened by the approach of a predator they make alarm calls, alerting others in their family group to the danger. Researchers deciphered their calls in 2013 and found that they have a consistent structure and convey different types of information. Using two types of call in combination, titis are able to indicate whether a threat is a raptor (such as an eagle) or a ground-based predator such as a capuchin monkey or oncilla (small spotted cat) and whether it is in the tree canopy or on the ground.

> **NOW OR NEVER**
>
> **There seems to be a window of opportunity for humans learning a first language that remains open during childhood, but then closes around the time the child starts to become a teenager. Feral children and those raised in isolation generally are not able to learn a language fully if they aren't exposed to speech before the age of 13 or so. This may be because the parts of the brain normally used to control speech have not developed or have been assigned to other functions. Those children who are found and brought into a social setting when very young can often learn to speak a language as competently as a child raised normally in a social setting.**

Talk to me

It's one thing to distinguish the features that encode meaning in the sounds an animal makes, but another to show that other

THE FORBIDDEN EXPERIMENT

The fascination with how we acquire language is long-standing and, in the past, people were less fastidious about how they studied it. The so-called 'forbidden experiment' involves raising a child without exposure to language or, sometimes, to any human company. It's been carried out several times – probably more than have been recorded – though without the rigour that a modern experiment would demand.

The Egyptian pharaoh Psammetichus I reputedly sent two infants away to be raised by goat-herds and suckled by goats; the goat-herd and his family were not allowed to speak with the children in any language. The Holy Roman Emperor Frederick II had children raised by nurses and servants who were forbidden to speak to them. Both rulers hoped that their experiment would reveal humans' first, natural language (Frederick suspected it might be Hebrew, but he was disappointed).

animals understand and respond to the sound. Many types of monkey, and other animals, call out when they find food. To test the amount of information there is in such calls for others to interpret takes some experimentation.

A group of researchers worked with bonobos at Twycross Zoo, England, to investigate how much meaning is communicated in the animals' food calls. Bonobos make five types of sound on discovering food, based on the extent to which they like the food. The researchers put popular food (kiwi fruit) or acceptable food (apple) in a bonobo enclosure and recorded the sounds the bonobos made on finding each. They later released a second set of bonobos into the same area (now vacated by the first set) and played back the recordings to see where and how enthusiastically the new set of bonobos foraged for food. Both troupes of bonobos were used to finding kiwi at one location and apple at the other.

The bonobos generally preferred to look at the kiwi site first, but this baseline preference was greatly increased by playing sounds of bonobos who had found kiwi fruits there. When the researchers played the sounds made by bonobos who had found apples at the apple site, the second troupe showed much greater interest than usual in visiting the apple site, and ignored the kiwi site. The results showed that at least some of the second group of bonobos were able to understand the information conveyed by the calls of the first group.

What's that dolphin called?

Dolphins make characteristic clicking and whistling sounds to communicate with one another. A research team based at the University of St Andrews, Scotland, tracked 200 bottle-nosed dolphins in the North Sea and eavesdropped on their conversations. They found that each dolphin has its own signature sound of clicks and whistles that includes information about its gender, location and state of health. A dolphin who wants to hook up with a previous close companion imitates the other dolphin's signature call, and a mother who has lost her offspring will call its signature sound until she finds it. The signature sound is effectively a name. Dolphins announce their name when joining a group, just as humans introduce themselves when they meet.

When researchers played back recordings of the signature sounds, each dolphin that heard its own name responded by repeating its sound – so you could call a dolphin register, if you had a lot of dolphins. The researchers found, too, that dolphins would move closer to an audio-speaker that played the signature sound of an individual already known to them.

Around half of all dolphin-speech comprises signature sounds; there is a lot of work still to be done to find out what they are saying in the other half.

Saying and singing

Birdsong sounds very different from human speech, but researchers have found that birds and humans use the same brain mechanism controlled by 55 of the same genes. The research consisted of sequencing the genomes of 48 types of songbird and analysing their brains, and looking at how birds learn their songs. Like young humans, baby songbirds first babble and stutter as they learn to 'speak'. Again, like humans, songbirds can be bilingual. Scientists hope that by studying songbirds they might be able to learn more about how humans develop speech and perhaps help treat speech disorders.

Besides songbirds, parrots and hummingbirds can learn to make new sounds. They sometimes put their skill to use imitating human speech. There are many examples of parrots and mynah birds that apparently use human language – but how much (if anything) do they understand of what they 'say'?

Who spoke first?

Birds have been around for much longer than human beings. They first appeared around 150 million years ago, and greatly diversified around 55 million years ago, whereas the earliest species of humans appeared only around 2 million years ago. We don't even know whether early humans used language.

Although the same genes are involved, this doesn't mean that humans and songbirds developed from the same

NOT JUST LANGUAGE

Birds don't limit their imitation skills to simply copying human speech. Parrots can learn to whistle, and even to copy the barking of a familiar dog. Some songbirds, such as starlings, also copy the noises made by inanimate objects including telephones, car alarms and chainsaws.

evolutionary ancestor. The last common ancestor of birds and humans lived around 310 million years ago. It is most likely to be an example of convergent evolution – the same solution emerging independently in unrelated organisms.

Words and signs

Language need not be speech or sound of any kind. People who can't speak may use sign language, something pre-vocal children can also learn to do. Research with primates has found that some of them can be taught to use sign language even though the vocal cords and brains of these animals apparently make it impossible for them to learn spoken language. (Their vocal cords do not close fully, and they don't have the musculature needed to control the tongue and lower jaw sufficiently for speech.) They can, however, use computer keyboards and signing to communicate.

Chimpanzees, bonobos, orangutans and gorillas have successfully been taught sign language. One chimp even spontaneously taught some of the signs to other chimps. There is little evidence that these primates have learned a language with a consistent grammar; it seems, rather, that they use the signs as unconnected symbols. But there have been some promising results in chimpanzees using American sign language (ASL).

Washoe was a chimp raised by a zoologist couple, Allen and Beatrix Gardner, in an environment as close as possible to that of a human child. She was taught around 350 ASL gestures and even combined them to create new terms, such as 'metal cup drink' for thermos beaker, and 'water bird' for swan. When one of Washoe's caretakers, who had been pregnant, signalled to Washoe that she had recently miscarried her baby, Washoe signalled back that she understood by drawing a finger down her cheek – the ASL sign for 'crying'. Chimps don't shed tears, but Washoe herself had lost two babies and appeared to register the pain of this in her empathic response.

Talking with animals

Many people feel certain their pets understand them and can communicate what they want. A dog bringing its lead to ask for a walk, or a cat standing by its food bowl is certainly communicating – but it's not using language. Language has a grammatical structure and words (or gestures) with different status related to their meaning – nouns, verbs and adjectives, for example. A dog that responds to the command 'Sit!' does not know the meaning of the word, it just knows that it will be rewarded with approval (and perhaps a treat) if it sits when it hears that sound. It is conditioning rather than understanding.

In the same way, an owner may come to associate a particular sound or gesture from their pet with something the pet wants. But they don't know the 'word' the sound 'means', they just know the pet wants food or to go outside. This is not language – the sound might mean 'please feed me' or 'I am hungry' – these are completely different sentences, but they elicit the same action from the pet owner. We are not 'speaking dog' when recognizing that a dog bringing its lead wants to go for a walk.

However, some animals can apparently learn enough human language to understand quite complex commands, even though they can't respond in words.

KOKO

Koko the gorilla has lived for 40 years, since the age of six months, embedded with humans in a research station in California. During that time, she has learned to communicate with humans using American Sign Language. Research published in 2015 revealed that video footage of Koko also shows she has learnt to control aspects of her breathing and vocalization which gorillas don't use in the wild. This includes coughing, blowing her nose and blowing a raspberry. The results suggest that given the right environmental triggers, some primates have more control of the body parts needed for speech than previously thought. It doesn't necessarily mean they could be taught to speak – if any gorilla was going to, Koko probably would have done so.

Kanzi, a 33-year-old bonobo, recognizes a lot of words in English. He can respond correctly even to commands that are not standard combinations of words he has not heard before. On one occasion, he carried a microwave oven outside when told to do so – this was not something he had done before.

Kanzi learned initially by eavesdropping on lessons which researchers were giving his mother. He began to use a system which allows him to put together rudimentary sentences using a keyboard with symbols (lexigrams) that stand for words. The computer then pronounces the sequences. The language system used is called 'Yerkish'. It's an artificial language with its own syntax, designed specifically for communicating with other primates. When asked to identify objects in a test with 180 trials, Kanzi gave the correct answer

93 per cent of the time; he answered complex questions with 74 per cent accuracy. Asked to 'make the dog bite the snake', Kanzi searched a selection of objects and chose a toy dog and a toy snake, then used his thumb and forefinger to close the dog's mouth around the snake, showing clear understanding of the request.

Birds are not bird-brained

Although primates can't vocalize human speech, some birds can do so. Parrots are the most famous examples, not only because they are able to repeat words and phrases, but also because they demonstrate some understanding of their meaning. The most famous of all talking parrots was Alex, a grey parrot trained by Dr Irene Pepperberg at the University of Arizona. Alex could respond to questions and say some phrases spontaneously. Scientists' interpretations of his intellectual ability and skill in communication varied; some felt he had not learned much but was responding to conditioning. However, he has the distinction of being the only animal ever known to ask a question. When learning colours, he asked, 'What colour am I?' Pepperberg repeated 'grey' six times and Alex then said, 'I'm grey'. This ability to pose a question rather than simply to answer one or respond to a command represents a very different form of communication. Unfortunately, Alex died at a relatively young age, so Pepperberg has had to begin her programme again with a new parrot.

OPERANT CONDITIONING

An animal can be trained to change its behaviour by being taught to respond to a stimulus – a punishment or reward. This is called operant conditioning. The first work on operant conditioning was carried out on cats by Edward Thorndike in the late 19th century. He found that cats learned to repeat a behaviour that brought a reward. Later work showed that animals learn to avoid behaviour which brings a punishment. Most communication with pets is operant conditioning in one form or another.

CHAPTER 14

What's happening with the climate?

Like it or not, believe it or not, the climate is changing. But that's nothing new – it has always been changing.

Long-range back-casting

It's much easier to tell what the weather has been like in the past than to predict what it will be like in the future. But looking backwards is a good preparation for looking forwards, as forecasting is rooted in understanding the conditions that have produced certain outcomes in the past and then trying to match current conditions to a set of past conditions.

What do we know?

Our records of the weather only stretch back a few hundred years. For some parts of the world, we have detailed and reliable records going back only around a hundred years, since the first meteorological organizations began collecting daily data. Prior to that, the only records are amateur observations and accounts in diaries, letters and chronicles – especially those describing unusual or extreme weather events. The longest-running continuous meteorological record in the world is the Central England Temperature series, covering an area of England extending from the south Midlands to Lancashire, which began in 1659.

CLIMATE AND WEATHER

Weather relates to short-term conditions: Is it raining? Is it foggy? Is it windy? Climate relates to longer-term patterns in the weather: Is the average temperature this decade higher than it was 50 years ago? Has average rainfall changed over the last three centuries?

While isolated weather events don't indicate climate change, a changed pattern of weather events can do so. So once-in-a-century floods are not significant. But if 'once-in-a-century' floods happen for six years out of ten, that probably is significant.

Of course, it's not possible to measure and record temperature unless you have a thermometer with a consistent scale. The first step towards this was the 'thermoscope' – forerunner of the modern thermometer – developed by the Italian scientist Santorio Santorio in 1612. He used this device to measure the temperature of medical patients. It was a simple glass tube, part-filled with water. Like all non-electronic thermometers, it worked on the principle that changes in temperature cause the air or liquid in a tube to expand or contract, causing the liquid in the tube to rise or fall. This was only a rough guide to temperature, as it lacked any scale.

The first thermometer with a proper temperature scale was developed around 1650 by Olaus Roemer, a Danish astronomer. His thermometer used wine as the liquid and set the boiling point of water at 60 and the melting point of ice at 7.5. It's difficult to imagine why he should have picked such an extraordinary scale. Daniel Fahrenheit improved the design by using mercury instead of wine or water, and made 0 degrees the freezing point of salt water. On his scale, 32 degrees marked the freezing point of pure water and 212 degrees the boiling point.

A more practical system – especially for scientific studies – was developed by the Swedish physicist Anders Celsius. It divides the temperature range between the freezing point and boiling point of water by 100 degrees, but Celsius made 0 degrees boiling point and 100 degrees the freezing point! It took the botanist Carl Linnaeus to swap them round to produce the scale in common use today.

Rainfall and snow are easier to measure than temperature. Even before thermometers, records often mention unusually cold or hot weather, drought, floods, fierce winds and so on. The Gaelic Irish Annals are some of the oldest chronicles in Europe and record events such as an extreme cold spell in the year 700 and gales in 892 which blew down trees and buildings.

The natural records

If we look outside human recordkeeping, we can often identify general weather conditions for a particular year by looking at tree rings. A slice through a tree trunk reveals its pattern of growth in the form of light and dark rings marking autumn/winter and spring/summer growth patterns. One ring represents each year of the tree's life. The width of the rings is a clue to the growing conditions that year. Extreme drought, for example, produces narrower rings because the tree is not able to grow as much as usual. The tree-ring record includes fossilized trees and provides a guide to weather conditions going back around 9,000 years in some parts of the world (mostly Europe and North America).

We can go back even further with the study of coral reefs and ice cores. Coral, like trees, grows in annual bands. The record can be a bit harder to read as the width of a band is affected by water clarity and the availability of nutrients as well as climate.

For evidence going back yet further, scientists take samples of Antarctic ice (see photo below), which also shows annual patterns reflecting climate, pollution and levels of gases in the atmosphere. The record reconstructed from ice core evidence goes back 800,000 years – quite long enough to compare temperatures and carbon dioxide levels with pre-industrial and virtually pre-human levels.

We can't go back any further than 800,000 years with the same degree of accuracy, but paleoclimatologists (scientists who investigate the prehistoric climate) have mapped out cycles of hot and cold climate lasting thousands of years. The climate alters over extremely long time periods because of changes in the Earth's orbit and in the composition of the atmosphere and the Sun itself.

Most interesting for humankind has been the temperature over the last 500 million years or so, since the evolution of land-based life. During that time, there have been ice ages and warm periods.

Ice age

There have been at least five major ice ages. The first lasted 300 million years, from 2,400 to 2,100 million years ago. The second lasted from 850 to 630 million years ago, and was the most severe, with ice spreading as far as the equator. This phenomenon has been called 'Snowball Earth' to describe how the planet would have looked at that time, as it's possible that virtually all land was covered by ice. When the ice finally melted, after nearly 200 million years, Earth became very warm. Around that time, life seems to have suddenly proliferated, with many new multi-celled, complex animals and plants appearing. Organisms left the sea to inhabit the land and rapidly proliferated. The temperature rose to around 15 °C higher than it is now.

After the next two ice ages, one 460–420 million years ago and the other 360–260 million years ago, came the rise of the dinosaurs. Their world was much warmer than ours. At the start of their 165-million-year reign, it was about 10 degrees warmer than it is now; although it dipped down to just a couple of degrees warmer, it was back up to 6 degrees hotter than it is now by the time of the dinosaurs' demise.

Over the last 65 million years, following an early high that occurred 50 million years ago, the temperature has gradually fallen, until around three million years ago, when it was about the same as it is now. But that was just the start of the next ice age – one we are still living through. If it doesn't feel cold enough to be an ice age, that's because we're in a warm period within it. Also, it's because we've known nothing else so don't have anything to compare it with. If we'd been here when it was 14 °C warmer, we'd soon spot the difference.

The current ice age began 2.58 million years ago. It's characterized by periods of warmth (as now) and glaciation (cold) in a cyclical pattern. At the start of the ice age, the temperature shifted every 40,000 years, but now it does so every 100,000 years. The last glacial period ended around 10,000 years ago, which suggests it should stay reasonably warm for a while – and only become really cold again in tens of thousands of years. The end of the last glacial period marked the point at which humans began to settle and farm instead of roaming the land, hunting and gathering. We are a warm-period civilization.

DEFINING AN ICE AGE

Meteorologists define an ice age as a period during which there is at least one continuous large ice sheet. At the moment we have them at both the North and South poles, but the one at the North is in danger of melting completely during the summer months. We'll still have one ice sheet, the Antarctic, so the ice age will continue – but for how much longer?

Getting warmer

The pattern of temperature change during the last 400,000 years suggests a rapid ascent from a cold period to a warm period, and then a gradual and rather jerky descent back into the cold. It's possible, too, that the current warm period is atypical in remaining relatively warm the whole time. The previous warm period seems to have been characterized by extreme cold snaps that lasted a few hundred years each. Where would we be now if, say, the entire period of the Roman Empire or the early Chinese emperors had been a period of extreme cold? Would we be back where we were in 1600? Or might we not have survived at all? Or would the human

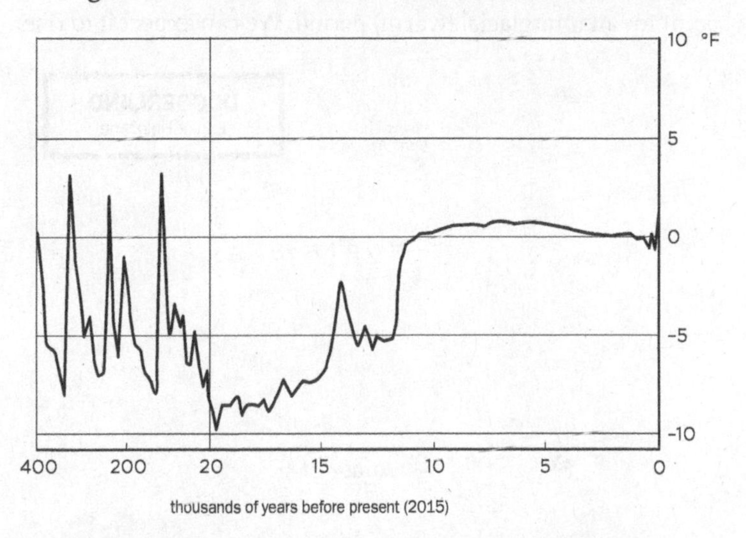

A graph showing the change in average global temperature over the last 400,000 years, with the last 20,000 years shown in more detail.

> ***'As global warming approaches and exceeds two degrees Celsius, there is a risk of triggering nonlinear tipping elements. Examples include the disintegration of the West Antarctic ice sheet leading to more rapid sea-level rise, or large-scale Amazon dieback drastically affecting ecosystems, rivers, agriculture, energy production, and livelihoods. This would further add to 21st-century global warming and impact entire continents.'***
>
> World Bank, 2012

population have declined to the levels of 70,000 years ago, with most of us concentrated near the equator in search of warmth? Perhaps we have been unusually lucky with the weather we have experienced so far.

So it might stay warm for another 50,000 years and get a good deal warmer before it grows colder. Whether or not climate change is caused by human behaviour, it is certainly happening. Following its normal trajectory, we can expect the average temperature to rise – it's nowhere near its usual high point for an interglacial (warm) period. We can expect it to rise

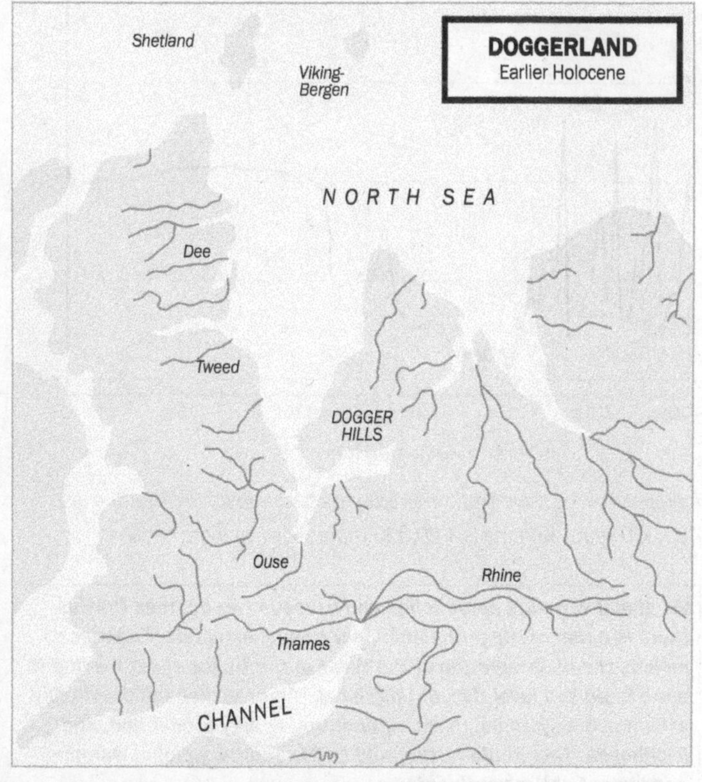

A map showing the hypothetical extent of Doggerland, which provided a land bridge between Britain and the rest of Europe around 12,000 years ago.

at some stage to around 5 degrees higher than present. With human-induced warming added in, it might rise much higher, much more quickly.

More or less land

As the climate warms and ice melts, the sea levels rise. When the Earth is at its coldest, sea levels are 120 m (390 ft) lower than they are now. Lots of water is trapped in ice. Whole tracts of land now underwater are revealed during these stages. During the last glacial period, humans could walk between Britain and continental Europe over a landmass called Doggerland, now sunk beneath the North Sea. (Its name is preserved, and familiar to sailors, as 'Dogger' – an area cited in the UK shipping forecast.)

Even a relatively small rise of a few metres will be enough to inundate low-lying islands; cities such as Venice, New York and Hong Kong will be at risk of flooding. This sea-level rise is certainly coming – the only question is whether it is coming in 100 years or in 15,000 years and therefore whether we have time to prepare for it.

And the weather?

A warming climate will bring different weather patterns, more extreme weather events, and different patterns of wind and sea currents. Many areas of the world are already seeing the pattern of weather change. Some places have hotter summers or more severe and frequent storms, or wetter winters.

The weather is an extremely complex system, affected not only by the temperature of the land, air and sea, but also by sea currents, solar activity, wind patterns and many other factors. The change of land use from forest to open farmland or cities also affects the weather, altering the behaviour of wind and cloud over land and the evaporation of water from the surface.

This contributes to the complex interchange of factors that determine the weather.

The weather system is chaotic. This doesn't mean there is no order to it, but that the order is very complicated and even a tiny change in one condition can have a considerable knock-on effect. Consequently, weather forecasting is notoriously difficult and unreliable, even in a relatively stable climate and over short periods. Forecasting the weather over longer periods is fraught with problems.

Hurricane Irma caused widespread destruction in the Caribbean and Florida Keys in September 2017.

All change?

Throughout Earth's history, plants and animals have had to adapt to changes in the environment. The species that adapt quickest and most successfully tend to survive; those that do not adapt often become extinct.

The natural world makes no distinction between changes in the environment brought about by human activity or those which occur through natural events. Some types of animals have already evolved in response to pressures originating with human behaviour (see *Are humans the pinnacle of evolution?* page 13). These changes have happened quickly, over the last few decades. Some plants and animals will adapt successfully to a changing climate, at least in the short term. If temperatures return to levels seen in the very distant past, it's likely that the natural world will look very different in the future. Many species alive today might cope with a rise in temperature of a few degrees, but some will perish and new ones will emerge.

THREATENED EXISTENCE

Polar bears rely on sea ice and a diet of seals to survive. As the seas become warmer, sea ice melts and it becomes harder for polar bears to hunt. Seals that live in very cold sea are pushed further north in search of colder water out of the polar bears' range. Polar bears might be able to adapt to eat different types of food, and/or change their behaviour to hunt mostly from land rather than ice.

Some polar bears are moving south, rather than north, and interbreeding with the closely related grizzly bears. This last change will still lead to the extinction of polar bears, but through their species changing slowly into another, not through mass starvation. This is how species have always changed and evolved. No one can tell whether the polar bears will fall victim to climate change or adapt to a warmer world.

Rising temperatures cause Arctic sea ice to melt, threatening the habitat and reducing the hunting grounds of polar bears.

It's not about us

Whatever the prospects for humanity, the planet and life in general have survived and adapted to many variations in climate. If the climate changes a great deal, humans as a species might not survive, but some form of life will. And as for the weather? It will follow the climate, but as for what it will be like next month or next year, that's still pretty much anyone's guess.

CHAPTER 15

Is this the end for antibiotics?

There are stories in the media about superbugs resistant to antibiotics. What has gone wrong with modern medicine?

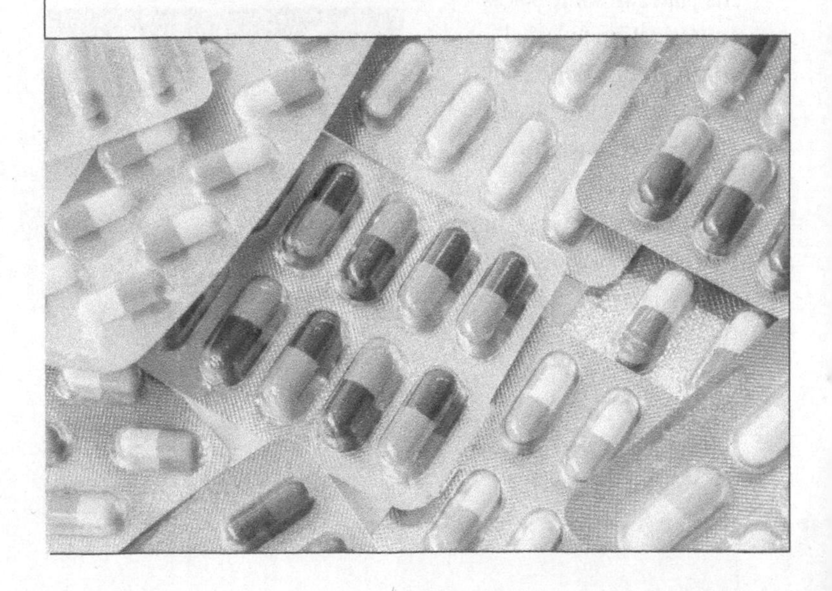

The bad old days

Up until the 20th century, there were no antibiotic medicines. People sometimes self-medicated, using substances known for centuries to have an antibacterial effect, but there was nothing like the powerful antibiotics available today. Although our bodies have an immune system which helps protect us from infection, many diseases have evolved to defeat the body's defence mechanisms. When this happens, the immune system can be catastrophically overwhelmed. In the past, people died of bacterial diseases and infected wounds which could be treated easily today with modern antibiotics.

Wonder drugs

The first modern antibiotic was penicillin, discovered by Alexander Fleming in 1928. He found it by accident, having gone away on holiday leaving behind a pile of unwashed glass plates on which he had been growing bacterial cultures. On his return, he found that something had cleared the bacteria on his plates in some places. Investigation revealed that a mould, *Penicillium notatum*, had produced a chemical that was toxic to the *Staphylococcus* bacterium he had been growing. In 1941, two pharmacologists, the Australian Howard Florey (1898–1968) and the German Ernst Chain (1906–79), developed Fleming's *Penicillium* extract into a useable medicine which we now know as penicillin. Its

use during World War II saved the lives of many injured soldiers who would otherwise have died of infected wounds. Fleming, Florey and Chain were awarded the Nobel prize for medicine for their work, estimated to have saved 82 million lives to date.

After penicillin, more antibiotics followed. Each one can only combat a particular range of infections caused by specific bacteria, so the search for new antibiotics continues. They seemed to be a wonder drug; suddenly, many previously fatal infections could be treated. Antibiotics are now so familiar to us it can seem astonishing that this happened only in the middle of the last century; there are many people alive today who recall the era before antibiotics.

Too good?

Antibiotics seemed too good to be true – and perhaps they were. It wasn't long before *Staphylococcus*, the bacterium Fleming had been growing, no longer responded to penicillin. Bacteria evolve quickly. They have such a short life cycle that they can cram lots of generations into a brief period, reducing the time it takes for useful genetic mutations to emerge which will help them overcome challenges or changes in their environment. They can also exchange beneficial genetic

MOSS AND HONEY

Honey and some types of moss have antibacterial properties and have been used to dress wounds for thousands of years. Honey works because it seals the wound, killing bacteria that need air to live, and its high sugar content causes bacterial cells to shrivel up. Some mosses contain a strong natural antibacterial. Soldiers who used sphagnum moss to dress their wounds, even as recently as World War I, found it very absorbent and it stopped the wound becoming infected.

mutations with other bacteria. So any mutation which allows a bacterium to resist an antibiotic will quickly spread to other bacteria. For the *Staphylococcus* bacterium, that challenge was penicillin. When *Staphylococcus aureus* became immune to penicillin, it should have been a wake-up call for medicine – but the warning wasn't heeded.

The use of antibiotics has increased astronomically. Not only are they used to treat infections in humans, they are routinely given to farm animals. When farmers found the yields increased if they fed cattle antibiotics, they began to do so on a massive scale, regardless of whether or not the animals were sick.

People also took antibiotics indiscriminately. They became accustomed to the idea that antibiotics were a cure-all and demanded them of doctors, who duly handed them out – even for viral infections against which they are completely ineffective. When patients failed to complete a course of antibiotics, or took them unnecessarily or unknowingly in tiny doses in meat and milk, the bacteria they targeted had a chance to build up resistance to the medicine. Exposure to a less-than-lethal dose of antibiotic enabled the bacteria to develop defences. An increasing number of bacteria have become resistant to the antibiotics most commonly used to combat them.

Bugs fight back

The rise of so-called superbugs such as MRSA began in the 1990s. MRSA stands for methicillin-resistant *Staphylococcus aureus* – it's a strain of the bacterium Fleming was growing. Methicillin was the antibiotic of last resort which was used to treat bacteria resistant to all other antibiotics. So when a strain appeared that was resistant to methicillin, it posed a major problem for doctors. Increasing the level of hygiene in hospitals has gone a long way towards fighting infection, and it's likely that cleanliness will, again, be more important in everyday life. If we can't fight an infection once it has taken hold, it's best not to let it start in the first place.

A losing battle

There are a few antibiotics of last resort which can still be used to treat superbugs resistant to our more routine antibiotics. But now even some of those are becoming ineffective and many medical researchers believe we are seeing the end of the antibiotic era. In 2009, scientists discovered an enzyme called NDM 1 that makes bacteria resistant to a whole class of our antibiotics of last resort. The gene that codes for this enzyme can pass easily between different species of bacteria, so sharing the resistance around. It's already widespread in India and has spread further afield, carried by travellers and in particular health tourists – those who travel specifically for medical treatment. The sexually transmitted disease gonorrhoea is already untreatable in some regions where an antibiotic-resistant strain is circulating. Drug-resistant tuberculosis is also on the increase, especially in Russia.

Antibiotic-resistant bacteria are part of a larger problem, the growing resistance of microbes to many antimicrobial medicines. This includes drugs used to treat fungal infections, parasite-borne diseases such as malaria, and viral infections including HIV.

Where next?

The world doesn't really have a plan for medicine in the post-antimicrobial era. Looking for further antibiotics might buy us a respite, but soon bacteria will become resistant to those, too. One possibility is to treat the bacteria to a bit of their own medicine and make them sick.

PIGGING OUT ON ANTIBIOTICS

In the USA, 80 per cent of the antibiotics sold are given to livestock. Feeding animals antibiotics means that farmers can keep them in unhealthy, crowded conditions without worrying about the animals falling sick and spreading disease between themselves. It's a practice that saves farmers money – but it might be very costly for the rest of us, if it means more and more antibiotics become ineffective at treating disease in humans.

Bacteria can suffer from infections just like any other organism. They can be infected by a type of virus called a macrophage, or 'phage'. Like other viruses, phages enter a cell, multiply, then burst out, destroying the cell and spreading into other cells. It might be possible to treat some bacterial infections by attacking the bacteria with specially chosen phages. Another possibility is using lysin, a type of enzyme that phages produce to break through the cell wall when they are ready to escape. Lysins applied outside the cell also seem to destroy it. It is important, however, to make sure only harmful bacteria are attacked and that cells which form part of the body, or bacteria which are beneficial to us, are left intact.

These treatments are a long way off becoming a mainstream option. In the meantime, we all need to use antibiotics thoughtfully and carefully to make them useful for as long as possible. We need to give research time to come up with the next miracle cure, before we completely exhaust the last one.

CHAPTER 16

Are stem cells the future of medicine?

There has been much speculation about whether stem cells can help treat many different ailments. But what are the facts?

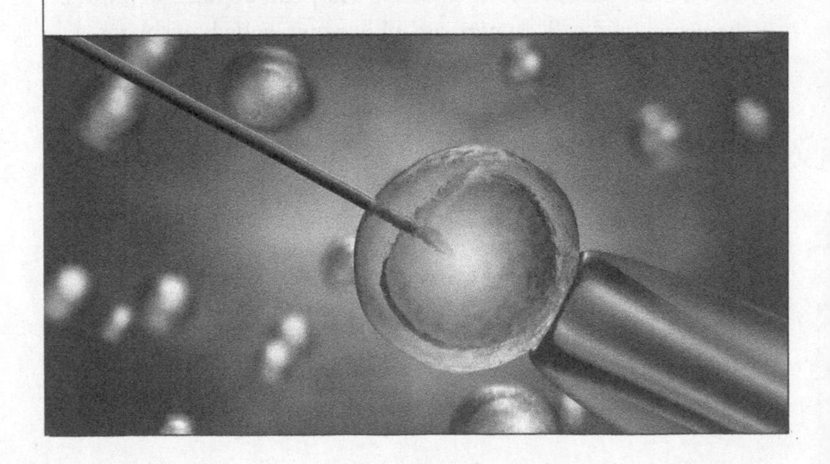

What are stem cells?

Stem cells get their name from the fact that other types of cells stem from them – they have the potential to develop into cells of different types. The cells that make up the very early embryo are called pluripotent stem cells (PSCs) and can turn into any type of cell. Later in the embryo's development, stem cells become more limited in the range of cells they can become.

After we are born we still have stem cells, but they are rather different and have even less potential. For example, deep within the skin are stem cells that can become different types of skin cell, but they can't become blood cells. In the bone marrow are stem cells that can become different types of blood cells, but they can't become skin cells. Stem cells are needed to repair damaged tissue, for growth, and to replace tissue that has a pattern of regular renewal such as the lining of the gut.

The uses of stem cells

Stem cells have been hailed as the way to repair damaged bodies, to treat diseases and maybe even to grow new tissues or organs.

At the moment, stem cells can be used to treat some types of cancers, blood disorders and immune deficiencies. Their use is often most successful in children. They can be used to produce healthy blood cells in patients whose own cells have been damaged by cancer treatments.

They can also be used in the screening and testing of new drugs, to produce tissue that can be manipulated and destroyed with impunity. Cell-level testing is an important stage of drug-testing. For example, stem cells could be triggered to produce liver cells for testing new drugs to treat liver disease. The liver cells would behave in the same way in the laboratory as they do in the body, so researchers could safely check the effects of a new medicine directly on liver cells before testing it on human patients.

KNOW YOUR STEM CELLS

There are different types of stem cell. Their names relate to their abilities, where they are found or how we obtain them.

- **Pluripotent stem cells** are the most useful. They are found in the early embryo and can produce cells of any type (from the same organism). So pluripotent cells could be used to generate blood, bone, nerve cells, liver cells, lung tissue, muscle, skin and so on.
- **Haematopoietic stem cells** are blood stem cells. They are found in the bone marrow and can grow into any of the many different types of blood cells we have. They can be used to treat blood disorders.
- **Embryonic stem cells** are derived from embryos. The single cell of the newly fertilized egg must quickly multiply in order to grow into all the different types of cells found in the body. Early embryonic cells are pluripotent – they can grow into anything. Since 1998, it's been possible to grow more embryonic stem cells from a sample.
- **Cord-blood stem cells** are taken from the umbilical cord when a baby is born. There is no harm to the infant in taking them. They are haematopoietic cells.
- **Somatic stem cells** are present in adults and children. They are tissue-specific and can turn into only a limited number of cell types. Skin stem cells, for instance, can produce the different types of cell needed to build skin.
- **Induced pluripotent stem cells** are cells derived from adults (or children) and returned to a state of pluripotency in the laboratory.

Treating leukemia

We have used stem cells for decades in the form of bone-marrow transplants to treat leukemia. New blood cells are created in bone marrow and released into the bloodstream.

STARTING FROM ONE

Most animals and plants begin as a single cell, called an egg cell. The egg cell splits repeatedly to produce more cells. After a short time, the cells start to take on special functions (differentiate), developing into different parts of the embryo. The first cells are capable of turning into any type of cell the organism needs.

Exactly how cells 'know' which type to grow into is not fully understood, but it is mainly controlled by chemicals released in the embryo. The cells take their instructions from DNA, which is in the form of long strands called chromosomes (see *What's the difference between a person and a lettuce?* page 207). DNA controls every living organism, from aardvarks to zebra fish.

Your DNA has a complete blueprint for your entire body, coded as a series of genes on chromosomes. Genes can be 'expressed', which means they are turned on and will have an effect on the functioning of the cell, or 'repressed', which means they are turned off and have no effect. By expressing the right genes in the right cells at the right time, the body grows and functions correctly. Bones develop where there should be bones and stop growing when they are complete; lungs grow where there should be lungs, and so on. Chemical signals tell a cell which type of cell it should become.

When a patient's own supply of new blood cells goes wrong, as it does in cases of leukemia, a bone marrow transplant from a close genetic match can correct this by providing a new batch of stem cells to create healthy blood cells.

Before the bone marrow transplant is given, the patient's own white blood cells must be killed off. As the white blood cells are key to our immune system, this leaves a brief and risky period in which the patient has no immune system – so no resistance to infection – before they are able to produce good, new blood cells.

Working with stem cells

Stem cells appear to know what needs to be done and then do it. If they are introduced to a body in need of urgent repair, the stem cells produce the type of cells that are needed – they can somehow detect damage and are programmed to fix it.

The nuts and bolts of growing stem cells

Pluripotent stem cells are taken from very early-stage embryos, up to 14 days after fertilization. At this stage, the embryo is a blastocyst (see illustration below), a mostly hollow ball of cells with a clump of stem cells at one end.

There are two sources of embryonic stem cells. Originally, they were taken from embryos created during fertility treatment but surplus to a couple's requirements. Typically, *in vitro* fertility treatment (IVF) consists of taking several eggs from the potential mother, fertilizing them, and then returning a small number to the woman's body to grow. More fertilized eggs are produced than are needed by the couple. Some might be frozen and stored for later pregnancies, but there is generally still a surplus. These might be donated for stem cell research.

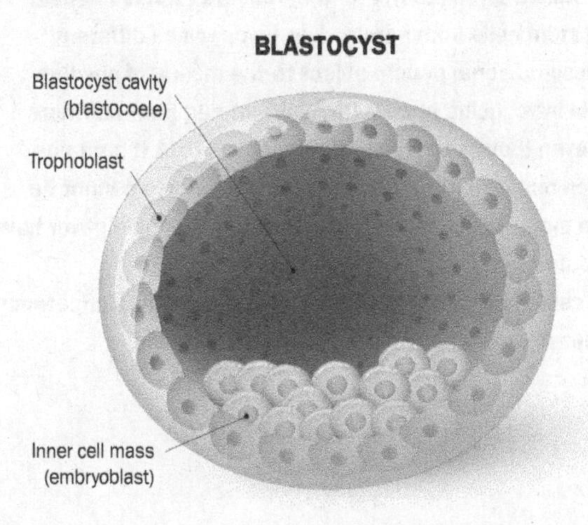

RIGHT OR WRONG?

The use of embryonic stem cells is fraught with ethical dilemmas and controversy. Human embryonic stem cells can currently only be harvested from human embryos, and the process involves destroying the embryo. Although the embryos used are surplus from IVF cycles, some people object that they could, if implanted, have grown into babies. The parents of the embryos do not want to have them implanted because their families are complete, so it is difficult to know what else to do with the surplus embryos if not used in medicine and research.

Using cells from cord blood has fewer ethical issues. There is no harm to the mother or infant in using this blood, which would otherwise be clinical waste. Some people want to store the cord blood for potential use by their child or another member of their family. In general, there is very little chance that the family will ever need it (unless there is already a known genetic condition that might benefit from treatment with it). Storing the stem cells from cord blood for personal use is unnecessary and over-cautious in most cases. Many hospitals with the facilities to take and store cord blood encourage parents to donate it to a cord-blood bank instead of paying to keep it. Their families will be given priority for a match later if it is needed.

Using stem cells from admixed embryos raises different ethical issues. Some people object to the mixing of species implied in introducing human DNA into an egg from another animal, even though the nucleus and all the DNA it contains have been removed from the egg cell. The embryo cannot be grown to maturity in any case, but they feel it should never have been created.

Adult cells that have been artificially returned to pluripotency are the least contentious form of stem cells to use.

Recently, scientists have discovered how to make human 'admixed' embryos. These are embryos made with the nucleus from a human cell (of pretty much any type) and an egg cell from another mammal. The nucleus, containing the DNA, is removed from the egg cell and the nucleus of the human cell is used to replace it. The cell is stimulated with electricity, which prompts it to begin dividing. The cells that grow are 99.9 per cent human, but have all grown from a single skin cell, or other common cell. It's illegal to implant admixed embryos, and they could never develop into a baby. Once the fertilized egg reaches the blastocyst stage, the stem cells are removed and grown on a culture medium. They reproduce rapidly, providing a supply of pluripotent stem cells for research or treatments.

A third method involves taking cells from human and reverting the cells to a pluripotent state – turning back the clock, as it were. These are called iPSCs, or induced pluripotent stem cells. Human iPSCs were first developed in 2007. The cells are reverted by using a virus to introduce chemical messages into them.

If they are found to be safe, there would be huge advantages to using reverted cells for therapies or research. There are no ethical issues, as no embryos are involved, and the original cells could be taken from the patient's own body, so there would be no issues with rejection of tissue grown from them.

New hope

Ongoing work with stem cells suggests they might be helpful for treating diseases in which certain types of cells have been damaged and destroyed. In conditions such as rheumatoid arthritis, immune cells attack the joints. Treatment with stem cells might involve removing the patient's immune cells (white blood cells) and replacing them with cells that don't attack the body. There are other possibilities, such as restoring eye tissue

Retrieving stored stem cell products from a cryogenic freezer in a laboratory.

after sight loss caused by macular degeneration, or treating Parkinson's or Alzheimer's disease by restoring brain tissue.

Stem cells could be used to build replacement tissue or organs outside the body for later transplant. Scientists can build a 'scaffold' – a basic structure for the organ – which is then populated with stem cells triggered to produce the appropriate type of somatic cells. Artificial bladders have already been created in this way. Despite the potential for life-altering treatments, scientists have to be cautious about stem cell developments. It's not clear exactly how stem cells differentiate or decide how far to multiply. Some experts fear that stem cells going into overdrive might cause cancer, which occurs when cell division gets out of control.

CHAPTER 17

How does a caterpillar turn into a butterfly?

A caterpillar spins a cocoon and after a few weeks a butterfly emerges. How does this happen?

All change

It is one of nature's miracles that an animal can completely transform from one state to another inside a cocoon. Many types of insect go through this metamorphosis. With frogs, we can even see the transformation happen: legs grow, the tail shrinks, the gills shrink and (one bit we can't see) the lungs develop. But what happens inside a cocoon (called holometabolism) is hidden from view.

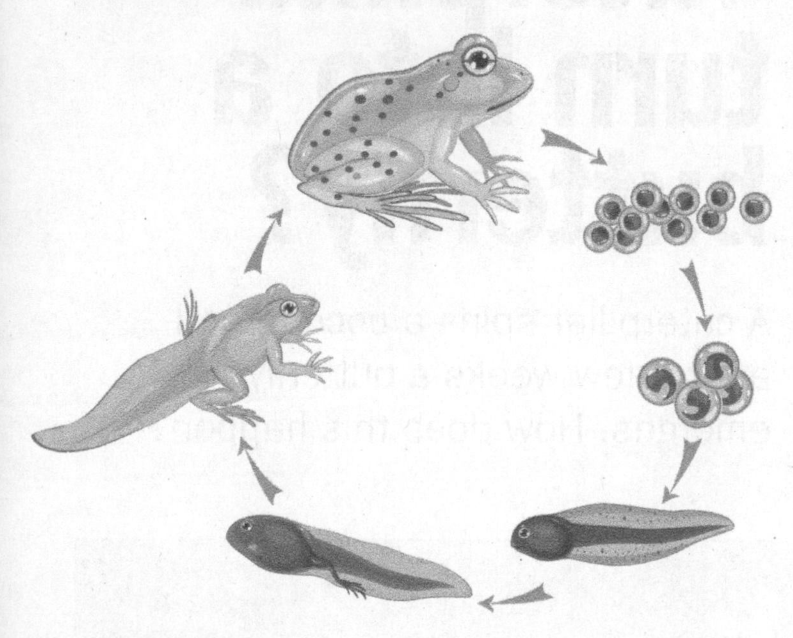

Insect stages

Insects begin life as eggs and hatch as grubs, maggots or caterpillars, which are all forms of larvae. The larvae feed and grow until they reach the point where they move on to the next stage of development – pupation. They then produce strands of liquid protein which harden in the air to form fibres that the larvae spin into a cocoon. The cocoon can be soft or hard depending on the species of insect. Some forms of cocoon fibre are called silk and can be made into textiles for human clothing.During the time the insect is in the cocoon it is vulnerable, because it can't move. Its best defence is to remain hidden. This is pretty much its only defence unless the cocoon is very hard or the insect itself is toxic to predators. Butterflies and moths often spin their cocoons somewhere out of sight, stuck to the underside of a leaf or hanging from the eaves of houses and sheds. Blowflies spin their cocoons in dark cracks and corners. Some moths even make their cocoons underground.

Inside out

Once the larva has spun its cocoon it is called a pupa and the whole caboodle – cocoon and pupa – is called a chrysalis. Interesting things begin to happen inside the chrysalis. First, the gut enzymes that up until now have been digesting the larva's food begin to digest its body from the inside out. At a certain point, cutting open a cocoon would reveal just a mushy caterpillar soup with no obvious body structure or parts at all. The biological miracle is that this mush transforms into a complete and complex insect. It is a phenomenon that only happens in insects.

Getting organized

The caterpillar's preparations begin even before it has hatched from the egg. At that stage, it forms highly organized groups called imaginal discs which are dedicated to specific structures in the adult insect's anatomy. There are imaginal discs for the eyes, wings, mouth parts, legs and so on. In many species of insect, the imaginal discs for the wings remain dormant throughout the larval stage, but in some they start to get to work early on. That means a few types of caterpillar and other grubs have vestigial wings already forming inside the body.

Ultimate transformers

Once the larva pupates, digestive enzymes break down most or all of the body cells of the pupa except for the imaginal discs and anything that has already developed from them. (In some species, some muscle and other tissue is also retained in the final insect.) Then the imaginal discs use all the caterpillar 'soup' inside the cocoon to build the cells needed for the organs and tissues of the butterfly (or whatever insect is brewing). There is enough material in the pupa to do this – it just needs to be completely reorganized. An imaginal disc that begins with only 50 cells might increase a thousand-fold over a couple of weeks to build an entire wing.

After a few weeks, or occasionally months, the final form of the insect, the butterfly or moth, emerges from the cocoon. It might bite its way through the cocoon or produce a chemical that softens it so it can push its way out.

SOUP REMEMBERS

A 2008 study involving moth caterpillars found that those trained to avoid a particular scent remembered their training as adult moths. The caterpillars were given a mild electric shock in association with the scent and quickly learned to avoid it, moving away from the scent in experimental apparatus. After metamorphosis, the adult moths also avoided the scent. This suggests that some parts of the central nervous system survive being dissolved into caterpillar soup and reconstructed.

CHAPTER 18

What is the most economical way to drive a car?

You'd think the faster you go, the more fuel you would use. But it's not that simple.

Anyone who has sat in a traffic jam knows that it takes lots of fuel to move very slowly. But why? It's all down to the amount of work the car engine has to do and how efficiently it does it.

Cars and cakes

An engine, like a human body, uses energy to work. Your body uses food as its fuel, breaking the bonds within molecules to release energy when you need to do something, whether it's to breathe, to heal an injury, grow your toenails or run for a bus.

A car engine does exactly the same, breaking the bonds within molecules to release the energy it uses – via the engine, crankshaft, drive shaft and axles – to turn the wheels, propelling the car along the road. Work is measured in newton-metres or joules. (They're the same thing – a joule is the amount of energy used to exert a force of one newton over a metre.) You may be more used to calories as a measure of the energy that is locked in food. A calorie is the amount of energy it takes to raise 1 g of water through 1 °C and is equivalent to 4.2 joules. We get our energy from our food; cars usually get it from petroleum or diesel fuel.

Energy, work and force

It takes energy to move your car along the road, and the energy is expended through exerting a force. Imagine that the car is not moving and you have to push it. You obviously have to exert a force to move it. Force is measured in newtons (N). A newton is the force needed to move 1 kg mass at an acceleration of 1 m per second per second. That means the first newton is spent accelerating the mass from standing still to moving at 1 m per second (m/s), and the second newton accelerates it to 2 m/s, and so on. Clearly this is not much force, so we usually measure forces in kilonewtons (kN).

If you have to push your car and it weighs 1,000 kg, you notice that it is very difficult to get it going, but once it's moving you don't need to push so hard. This is because you have to exert the most force to accelerate it, rather than to move it. You have to accelerate it from zero (standing still) to some speed (moving). To start it moving from stationary to 1 m/s takes 1,000 N (one newton for every kilogram of car).

Some laws (not traffic laws)

There are two important laws in working out how much fuel your car uses. The first is Isaac Newton's second law of motion:

$F = m \times a$

(Force in Newtons = mass in kilograms × acceleration in metres per second per second.)

The other is the first law of thermodynamics, often thought of as the rule of conservation of energy. This states that energy is neither created nor destroyed, but can change, for example, from heat to work (such as mechanical energy) or from work to heat.

The energy locked in fuel is chemical energy. When the fuel is burned, the chemical bonds are broken and the energy is released. The energy released from fuel is converted to other forms in your car, but not all of it is used to move the car along. It is converted to mechanical energy to drive the car, with heat as a by-product.

The problem with friction

There's another law that's relevant, too, and that's Newton's first law of motion:

'Every object in a state of uniform motion tends to remain in that state of motion unless an external force is applied to it.'

This means that if we set something moving, it will carry on doing so unless something acts to stop it. In outer space, a craft set in motion will keep going almost indefinitely because there is very little acting on it to stop it or change its path, unless it strays into the gravitational field of some star, planet or other body. The same is not true of an object on Earth, as there is always matter in contact with the moving object, whether it is air, water or the ground (or a combination of these). This means that there is always friction between the object and whatever it is touching. Friction with air or water is experienced as drag.

A moving car is subject to drag from the air it passes through and friction from between the road surface and the tyres. It takes work for the car to overcome the drag of the air and the friction with the road. With perfectly smooth tyres and a very smooth road surface, there would be less friction. But we need the friction because it gives the car good road-holding. Bald tyres on an icy road have little friction and are notoriously dangerous. So we are stuck with the friction, even though it means the car has to do more work, and that means it needs more fuel.

There is also friction between parts of the engine that are in contact with one another, so we use oil and grease to lubricate them, thereby reducing friction. Keeping the car well-oiled reduces the friction, so reduces the amount of energy used overall. That, in turn, reduces the amount of fuel the car needs to cover the same distance.

Imagine for a moment that you didn't need to worry about friction – that your car could glide along the road like a spaceship through space. Once the car is moving, nothing is going to stop it, so it doesn't need fuel to travel along. Fuel is needed only to get it moving at the speed you want it to go at – which returns us to the second law of motion:

$F = m \times a$

Here, it's clear that the faster you travel, the more fuel you will use because F (force) simply increases with a (acceleration). Once you are going at the desired speed – in this ideal frictionless world – it doesn't take any more fuel to keep going.

In our imperfect, frictionist world, we need to burn fuel to reach the desired speed, and then use more fuel to keep going, fighting friction with the road, within the engine, and drag. This is why the calculation becomes difficult.

Efficiency of movement within the engine is key. The most economical driving speed is not really a speed, but an engine state. For most cars, changing gear when the engine revs reach 2,500 for petrol engines (2,000 for diesel engines) gives the most economical ride. It's important, too, to accelerate and decelerate gently – revving the engine burns a lot of fuel, whether you are doing it to accelerate quickly, or because you are in too low a gear.

So the best speed is . . .

Fuel efficiency drops off rapidly at low speeds and at very high speeds. Overall, the most economical cruising speed is around 55 mph. Research by *What Car?* shows that travelling at 80 mph uses up to 25 per cent more fuel than travelling at 70 mph. At high speeds, air resistance increases rapidly, so the car needs to use much more fuel to maintain the speed. The power needed to push an object through a fluid (including air) increases as the cube of the velocity: if you double the speed, the force needed to counteract air resistance increases eight-fold (2^3). Doubling the speed also means halving the travelling time. Power is the rate of doing work, so total power is time-dependent. Therefore, doubling the speed increases the work to overcome air resistance eight-fold, but reduces the time by half, so the total power needed increases four-fold. Taking account only of air resistance, it takes twice as much fuel to travel at 100 mph as to travel at 50 mph.

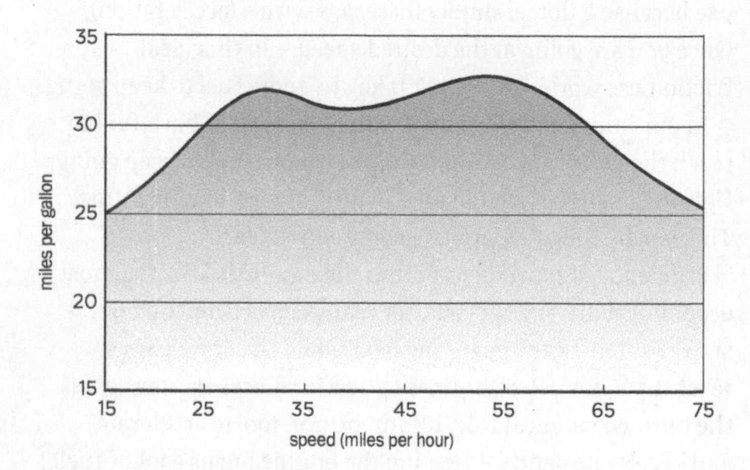

Can we trust manufacturers' claims?

In a word, no. The official figures for fuel consumption are based on tests that don't involve driving the car in typical conditions on a real road. The test-drive doesn't cover poor road surfaces, steering around hazards, accelerating aggressively at traffic lights or braking abruptly to avoid collisions. Instead, staid and sensible people drive the car around a virtually empty circuit at gentle speeds with slow acceleration and controlled braking.

In 2014, tests of 500 cars found the average fuel consumption was considerably greater than that advertised by manufacturers, giving 18 per cent fewer miles per gallon than

PHOTON SAILS

With friction removed from the equation, Newton's first law of motion can (at least in theory) be demonstrated by photon sails. The idea is that a spaceship can have a sail – like a yacht. As photons from sunlight (or another star) strike the sail, they have the same action as wind and cause the spaceship to move. As there is nothing to stop it, it takes only the slightest pressure from photons to get it moving.

they claimed. This will come as no surprise to anyone who has been disappointed by the fuel consumption of their new car. What is more surprising, perhaps, is that the fuel economy is worst with smaller cars.

As the chart below shows, it's better in terms of genuine fuel economy to drive a car with a 1–2 litre engine than one with an engine under 1 or over 2 litres.

Engine size	Claimed mpg	Actual mpg	% difference
Up to 1 litre	60.3 mpg	38.6 mpg	36 per cent
1-2 litre	59.1 mpg	46.7 mpg	21 per cent
2-3 litre	52.9 mpg	45 mpg	15 per cent

Are super-fuels really that super?

Fuel is rated by a research octane number (RON) that relates to how much the fuel can be compressed before it ignites. Basic unleaded fuel has a RON of 95, and super-unleaded has a RON of 98. A top-of-the-range fuel has a RON of 102.

In theory, the higher RON means that the car can get more energy – and therefore more mileage or speed – from the same amount of fuel. In practice, you are unlikely to get sufficient improved performance in most cars to justify the extra cost.

Super-unleaded fuels are most likely to bring a benefit to high-performance cars driven hard and fast. Turbocharged and supercharged engines work at higher temperatures and pressures than regular engines and can make the most of the higher octane-rating. They are also more susceptible to 'knocking' caused by unburned fuel pre-igniting in the wrong place and potentially damaging engine components. Higher-octane fuels are less likely to produce knocking, so can protect a supercharged engine.

HOW TO CUT YOUR FUEL CONSUMPTION

To save fuel, money and the planet, you can reduce the amount of work your car has to do in several ways:

- **Reduce weight: in the equation $F = m \times a$, m is mass; increased mass increases the force needed to move the car. Don't carry around weight you don't need, so don't leave things in the car you aren't going to use – if there are no toddlers with you, don't take their buggy; if it's not winter, don't take a snow-shovel.**
- **Reduce drag: remove things that stick out and spoil the streamlined shape of your car – so no roof-racks or bike-carriers unless you are actually using them.**
- **Keep your car healthy: have your car checked frequently so it runs as well as possible – keep the tyres pumped up, and everything well-oiled and smooth to cut down the work the engine has to do keeping things going.**
- **In hot weather, turn the air-conditioning off and open a window instead, at least at low speeds. At very high speeds, though, the increased drag of an open window negates the effect, so you might as well turn on the air-conditioning.**

TO BUY, OR NOT TO BUY?

Are you getting the mileage you paid for? Here's how to tell:

Super-fuel mileage	=		Super-fuel cost	=	
Normal mileage			Normal cost		
40 mpg	=	1.14	£1.30	=	1.08%
35 mpg			£1.20		

The mileage with super-fuel is 114 per cent of the normal mileage. So the cost of super-fuel is 108 per cent the cost of normal fuel. As this driver is getting an extra 14 per cent mileage for an extra 8 per cent cost, it's worth buying super-fuel.

On balance, if you have a high-spec, powerful car with a turbocharged engine, you might get value from a super-fuel. If you have a run-around, or most of your journeys are short and traffic-clogged, super-fuels are likely to be a waste of money. If you want to try them, fill the tank and note the mileage you get from it – perhaps do this three or four times to get a good average – and then compare it with your usual miles-per-gallon.

CHAPTER 19

Why do we find seashell fossils on mountains?

Fossils of sea creatures are sometimes found in places very far from the ocean.

Lots of fossils

Around 99.9999 per cent of all the species that have ever lived are now extinct. Only a tiny proportion have ever been fossilized, but that still represents a massive number of fossils. Most fossils are not of huge, imposing creatures like dinosaurs; they are small, even microscopic fossils of plants, sea creatures, insects and so on. Although there are large deposits where many spectacular fossils are found, it is far more usual to find smaller deposits and isolated fossils.

Continental drift

Animals and plants have changed over time, as have the land, sea and climate. It's possible to find fossils of sea creatures far inland because the areas they lived in long ago, at the coast or at the bottom of the sea, are no longer where they once were.

In 1620, Sir Francis Bacon noticed that the coast of West Africa fitted quite neatly against the Atlantic coast of South and North America. It was the first clue that the landmasses might not always have been where they were. The first person to propose that the continental landmasses actually move was a German meteorologist and geophysicist called Alfred Wegener (1880–1930) in 1912. He pointed to the existence of identical rock strata in South Africa and south-east Brazil, and to the fossils of the dinosaur Mesosaurus, found on both continents. The existence of coal in both Britain and Antarctica was further evidence. Coal forms from dead trees, but only in hot, wet conditions. Neither Antarctica nor Britain has a climate that could produce coal today.

There were only two possibilities: either the climate in these places had once been very different, or the landmasses had moved and had once been closer to the equator. Wegener felt that unless the Earth's orbit around the Sun had changed, there was no way Antarctica could ever have been warm

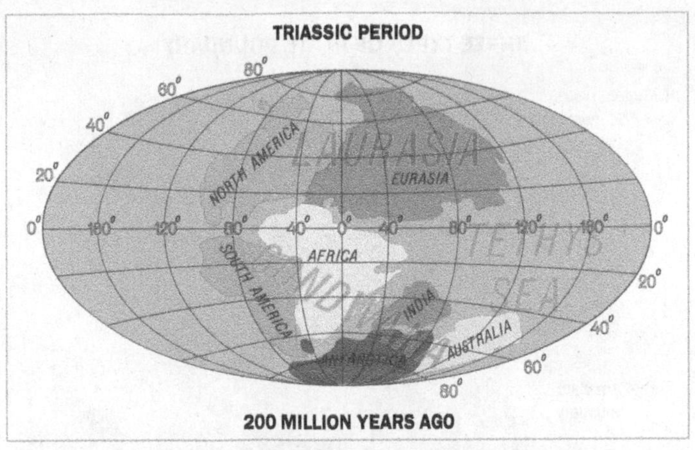

The location of the continental landmasses 200 million years ago.

enough for the formation of coal, so the answer must be that the landmasses had moved. Although he presented plenty of evidence to support his idea that the land on Earth has moved around during Earth's long history, Wegener couldn't say how it did so. As a consequence, the idea was fiercely opposed by the geological establishment when he first proposed it. But, during the 20th century, further evidence emerged that led eventually to the explanation – plate tectonics.

The Earth has a thin, rocky crust which sits on top of thicker layers of semi-molten rock, called magma. Convection currents produced by uneven heating from the Earth's core cause the magma to move, dragging the crust along with it. The crust is divided into seven large sheets, called plates, and several smaller ones. The boundaries between the plates are important locations in terms of Earth's geology.

New rock for old

In some places, the plates slowly move apart as magma pushes up through the gap from below and hardens into new rock. This happens at the mid-ocean ridges – beneath

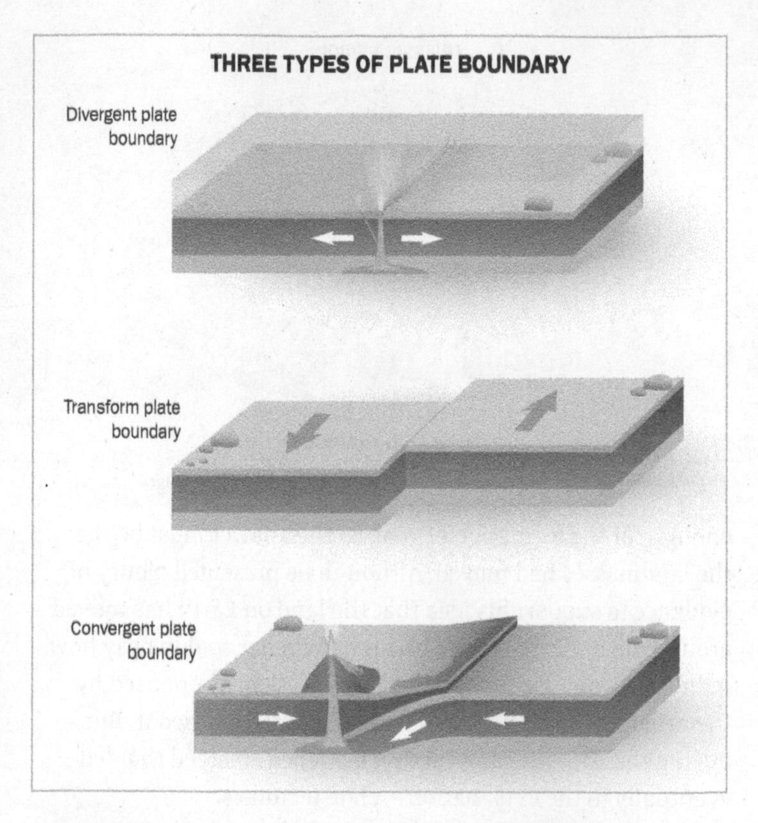

the Atlantic Ocean, for example. The new rock often forms underwater mountain ranges (the 'ridge') with a rift in the middle. The rift is the productive part where the magma is streaming out. This is what causes the volcanic activity in Iceland and other places, where magma erupting from undersea volcanoes forms new islands. Iceland is at the end of the Atlantic ridge. As the plates are pushed apart by rising magma, they move towards and push against the plates carrying the continental landmasses.

In other places, two continental plates may simply slide past each other. Earthquakes are common at these points, as pressure builds up when the plates become wedged against

HOW FOSSILS FORM

There are different types of fossils. The most famous, such as the great dinosaur skeletons we see in musuems, form when the hard parts of a dead animal or plant slowly turn to stone through a long process of chemical exchange. This only happens in exactly the right conditions. Occasionally, non-bony parts such as scaly skin can also be fossilized, but this is less common.

If the plant or animal is absent and only the impression of it is left, this is known as a trace fossil. Often made in mud or sediment that then turns to stone, trace fossils include footprints, tail drag marks or hollows made by worms and other burrowing creatures.

Fields of fossils can form when a local disaster such as a mudslide rapidly buries a large number of plants and animals at the same time. They are compressed and hidden (they can't be eaten by scavengers, blown away or trampled) and are slowly fossilized over thousands or millions of years.

Prehistoric sea creatures fossilized in rock.

each other until finally they jolt into a new position. This jolt causes an earthquake. The San Andreas Fault in North America is an example of a transform boundary where plates grate against each other, causing earthquakes.

At other plate boundaries, called subduction zones, the edge of the oceanic plate containing seabed sinks beneath a continental plate carrying the landmass. The seabed contains a lot of seawater that reduces the plate's melting point as it is forced down into the molten rock below. The rock of the seabed melts, forming magma that then squeezes up through gaps in the continental rock above to form volcanoes. Earthquakes and volcanoes are common in places like the western coast of South America, close to the subduction zone where oceanic and continental plates meet.

Forming mountains

When two continental plates meet, the edges crash into each other and push upwards, forming mountain ranges. The edges of the plates were once coastline, but become locked in land when plates collide. The Himalayas formed as the tectonic plate carrying India slowly pushed into the plate carrying Asia. This process is continuing, so the Himalayas

A tourist takes a photo of the crack between tectonic plates in Pingvellir national park, Iceland.

are growing a little higher every year. At the same time as growing, mountains are worn away by the weathering effect of wind and rain (called erosion). If they erode more slowly than rock is pushed upwards, they continue to grow. If erosion is more rapid than growth, they are slowly worn away. Once mountains are no longer growing, they slowly reduce in size as the weathering effect of wind and rain gradually wears them away.

The movement of the tectonic plates explains how fossils of sea creatures get to be on high mountains such as the Himalayas. Over millions of years, land that was once the beach of an island can become the middle of a mountain range. The rock that is now part of Mount Everest was once the coast of Asia and the coast of the island of India. The rock has carried its fossils of sea-dwelling creatures with it even as it has moved very far from the coast.

Still going

The plates have not stopped moving and are not likely to stop soon, though they are slowing down very gradually as the Earth cools. They are still moving at a rate of a few centimetres every year, though different research groups give slightly different figures for the exact speed. It seems that the plates are moving in the Atlantic at a rate of about 1 cm (0.4 in) a year each, causing Europe and North America to move apart and the Atlantic Ocean to grow at a rate of about 2 cm (0.8 in) each year. In a million years, it will have grown 20 km (12.5 miles). In some other places, the rate of movement could be 5–10 cm (2–4 in) a year.

If we could zoom through time a few hundred million years, we might see that the Pacific Ocean has closed up and the Atlantic vastly widened. Fossils of fish that die on the Asian coasts of the Pacific, and creatures from the beaches of California, might be high up in a new mountain range, perhaps joining the east coast of China with the west coast of North America.

CHAPTER 20

Can plants feel pain?

We know that other animals can feel pain. But what if it turned out that plants can suffer too?

More to them than meets the eye?

Plants make up a large proportion of the world's biomass, yet we know remarkably little about how they work. What has become increasingly evident over the last 20 years or so is that they are much more complicated than we have suspected. Because they are largely immobile and many of their responses are very slow (achieved through changing growth patterns), their active lives have gone largely unnoticed.

The ethylene 'scream'

In 2002, researchers in Bonn, Germany, announced an experiment which was widely reported as demonstrating that plants scream or whimper when damaged or sick. This sounds very dramatic, as though plants actually 'feel' pain as an unpleasant sensation, the way we do. The sound of the plants 'screaming' was picked up using a laser microphone. To say that plants scream is a bit of a misrepresentation, though.

Plants emit the gas ethylene in large quantities when they are damaged. The researchers in Bonn trapped the ethylene and bombarded it with an infrared laser, making the molecules vibrate. The laser beam was interrupted 2,000 times a second, making a high-frequency pulse. As the ethylene molecules were excited by the laser beam, they emitted a tiny amount of energy that was trapped and delivered through a resonance tube so it emerged as a sound. The researchers recorded the sound. They found that the more stress the plant is subjected to, the louder it 'screams' – but this is really just a measure of the amount of ethylene it releases.

The burst of ethylene is converted to the sound of a scream by the experimental equipment. It could as easily have been converted into a flash of light or a burst of heat. Using the same equipment, healthy plants emit a bubbling or gurgling noise as they produce ethylene in a more measured way.

The question of why the plants produce large quantities of ethylene when they are harmed remains to be answered. It could be that the ethylene is intended to deter attackers – insects or herbivorous animals – that are trying to eat the plant. Or it might act as a warning signal to other plants. This might sound far-fetched, but plants are capable of a lot more communication than we might imagine. If plants 'talk', they talk with chemicals. The ethylene scream is only the tip of the iceberg in terms of plants' chemical signalling and communication.

Ouch!

Before we go any further, it's worth pausing to think about what 'feeling pain' involves. If you touch something hot, two things happen in your body. Sensors in your skin called nociceptors detect the stimulus (called nociception) which indicates damage to your body and sends a message to your central nervous system (spinal cord and brain). The first thing that happens is an immediate reflex action to withdraw from the hot object. You don't have to think about it, and the signal doesn't even need to reach your brain – it can be handled in the spinal column in a 'reflex arc'. Your central nervous system sends a signal by return to the muscles in the arm to make

VEGETABLE NOISE

The research into the ethylene 'scream' has practical applications; it is useful in agriculture and retail. The 'sounds' from fruit, vegetables and plants can determine whether they are healthy or not, revealing disease before it is visible. This is valuable information for farmers, horticulturalists and retailers. Noises from cucumbers, for example, have revealed mildew before it can be seen.

you snatch your hand away from the heat and protect you from further damage.

At almost the same time as you move your hand, your brain receives the signals from all the nerves in the hurt area and collates them, determining the level of damage and pain. It's this part that gives the sense of being hurt and the knowledge you are in pain. The process is slightly slower than the defensive reflex action. You might have noticed there is a moment when you have seen you have hurt yourself but don't yet feel pain; that's the time your brain is taking to process the information. Pain is necessary as part of the learning process. It teaches you to avoid that dangerous situation in future.

A wide range of animals demonstrates nociception – from tigers to sea slugs. Feeling pain, however, is handled in the outer part of the brain, called the cortex. This is highly developed in humans. It is less well developed in other mammals, smaller again (proportional to overall brain size) in birds, reptiles and amphibians, and smallest of all in fish. Invertebrates don't have a cortex because they don't have a brain with the same structure as a vertebrate's brain. But we can't rule out the possibility that they have another way of sensing pain.

Plants don't have a central nervous system or pain receptors at all. How, then, are they sensing damage and responding to it? Whether or not plants 'feel' pain in any way comparable with the way we do, they are certainly a lot more sophisticated than we realize.

Plant 'senses'

Plants are very different from animals. They don't have obvious behaviours in the way animals do, they don't make a noise (that we can hear) and they tend to stay in one place. But this doesn't mean they aren't doing anything. They still respond to external stimuli – indeed, they respond to many more types of stimuli than we do.

Humans have five basic senses – sight, hearing, taste, smell and touch – but plants have many more if we measure sense by the ability to detect and respond to stimuli. Plants respond to heat, light, gravity, water, soil structure, nutrients, toxins, microbes, predatory animals and insects, and chemical signals released by other plants. Mouse-ear cress can respond to a magnetic field, and young poplar trees can detect if they are vertical or tilted. Plants have even been found to respond to touch and sound.

These detection mechanisms are generally called 'tropisms' in plants, rather than senses. Plants are phototropic – they grow towards the light. They are also geotropic: roots grow towards the ground, where gravity is strongest, and shoots grow away from it, against gravity. Investigating plant responses generally involves measuring chemical changes, electrical signals, and watching slower responses such as growth patterns.

Some plant responses are quite extraordinary. One study found that plant roots would grow towards an underground pipe carrying water, even though the outside of the pipe was completely dry. Plant roots approaching an impenetrable obstacle, such as concrete, divert before reaching it. They also avoid toxins and the roots of strong competitors by growing away from them.

Recent experiments with plants have uncovered more surprising results than the ethylene scream.

Australian research scientist Monica Gagliano carried out an experiment using mimosa, a plant that briefly collapses its leaves when disturbed. She dropped the plants 60 times at intervals of five seconds, from a height of 15 cm (6 in), catching them without harm each time. After five or six drops, many of the plants stopped collapsing their leaves, having apparently categorized the drop as non-threatening. By the end of the training period of 60 drops, all the plants kept their leaves fully open. It might be too anthropomorphizing to say that the plants 'learned' that the drop was harmless. To make sure their response was still functional, Gagliano then shook the plants. They collapsed their leaves with this new stimulus, so were clearly still able to do so. Gagliano retested the 'taught' plants every week and found they retained the lesson (that is, they didn't respond to a drop) for at least four weeks. Insects have a far shorter attention span, forgetting such a lesson after a couple of days.

CO-OPERATIVE BEHAVIOUR

It seems that plants have some kind of kinship recognition, too. When closely-related sea-rocket plants were put in the same pot, they shared resources cooperatively whereas usually plants tend to compete with one another in such a situation.

Carnivorous plants

Further evidence of plants processing information in ways we don't yet understand comes from the strange world of carnivorous plants. These gain some of their nourishment by consuming small organisms. They normally live in nitrogen-poor environments, so trapping and digesting insects and small animals provides the missing nutrients.

'Plants have both short- and long-term electrical signalling, and they use some neurotransmitter-like chemicals as chemical signals. But the mechanisms are quite different from those of true nervous systems.'

Lincoln Taiz, emeritus professor of plant physiology at University of California, Santa Cruz

There are many carnivorous plants, but the most fascinating is the Venus flytrap, because it moves. It has specially adapted leaves arranged as 'traps'. They have tiny trigger hairs which respond to an insect walking into the trap. When this happens, the trap closes and the plant secretes digestive enzymes which kill and dissolve the insect. The plant then absorbs the nutrients.

The plant will only close if two trigger hairs are touched within a period of 20 seconds. Even then, it only closes partially at first. It will close completely and begin the digestive process if the trapped item continues to touch the trigger hairs. This prevents the plant wasting effort trying to digest a bit of debris that has blown into it, or a pencil that an inquisitive person has poked it with. It also allows tiny insects to escape – there is too little nutritional value in them to justify the expenditure in digestive juices.

So how does the trap work? According to Alexander Volkov, professor of chemistry at Oakwood University, Alabama, it is formed from two lobes of a single leaf, with a hinge in the middle. The hairs are mechanosensors which convert mechanical energy into electrical energy. When an insect brushes against the hairs it triggers an electrical charge that opens specialized pores in the outermost layer of the trap's cells, allowing water to rush from cells on the inside of the lobes to cells on the outside. The dramatic change in cell pressure causes the lobes to snap shut on either side of the trap's hinge.

But how does it time the 20 seconds, 'remembering' that it's been touched once? No one knows. Here's a plant that has some way of storing information, but without using an animal-like nervous system or musculature.

Using pain

Animals use pain as a signal to get out of harm's way. If an organism cannot take evasive action, what use is pain?

Plants' response to damage can help to protect the injured plant and even other plants in the immediate area. They release chemical signals in response to some stimuli; these are carried through the air and picked up by other plants, which then respond. When a plant is attacked by insects or grazing animals, it releases a chemical signal which prompts nearby plants to produce chemicals that make them unappealing or even poisonous.

Injured tomato plants, for example, produce a chemical called methyl-jasmonate. This deters insects that might be feeding on the plant, but it is also detected by other plants, which then begin their own defence procedures, changing their chemical composition to produce chemical protection. It even works between species of plants in a wonderfully cooperative defence system.

Many plants produce toxic or otherwise repellent chemicals when they are attacked by insects, but some even marshal their defences when an insect lays eggs on their leaves, preparing in advance for the hatching of hungry caterpillars. Some plants can distinguish between mechanical damage – such as that caused by cutting the plant with a knife – and an attack by herbivores, responding to chemicals in saliva.

Working together

Chemical signalling happens not only through the air, but also through the ground.

Trees in a forest are connected by a massive underground network of fungi that grows in and around their roots. Some biologists have called it the 'wood-wide web'. By means of the fungi, trees pass chemical signals – and information – among themselves. But they also pass food around, even between species. Research in Canada has found that larger trees help out smaller ones, sending them nutrients such as carbon while they are overshadowed and less able to photosynthesize. An experiment which involved injecting radioactively-tagged carbon into fir trees found the carbon rapidly spread between all trees in the area. Evergreen fir trees were found to be helping deciduous birch trees, supplying them with nutrients during the winter when they couldn't photosynthesize and taking carbon (as glucose) back from them in summer. Dying trees will even dump their carbon to be carried by fungi to other, healthy trees.

Research with tomato plants and beans has found that those allowed to share root-fungi (called mycorrhizal fungi) also pass on information about attacks, preparing other plants to defend themselves. When one of a pair of linked tomato plants was infected with blight, the second plant became blight-resistant. And when one of a pair of fungally-linked bean plants was attacked by aphids, the second plant increased its aphid-deterrence chemicals.

ADAPTATION

One study found that playing a recording of caterpillars eating leaves to a plant of the same species of the victim primed it to produce chemicals that repel caterpillar attacks.

The dark web

The web can be used for crime as well; some plants steal from their neighbours. The phantom orchid has no green parts and can't photosynthesize, but steals the carbon it needs from nearby trees, using the fungal network. And some are guilty of even more substantial crimes. Marigolds and the black walnut tree produce toxins that can be carried by fungi to poison other plants which might want to share the space, competing for water, nutrients and sunlight.

Smart plants

Plants, then, are gathering information about their external environment and responding to it in a way that looks remarkably like a choice. But plants don't gather, process and pass on information in the same ways as animals. Even committed plant neurobiologists are not looking for nerves and brains in plants.

The 'choice' is biochemically mediated – but, much as we might dislike the thought, so are our own choices. Our brain biochemistry is as intractable as the plant's sensory mechanisms. One suggestion is that plant 'intelligence' might be something like the swarm or hive intelligence demonstrated by social insects such as ants.

A brain would not be useful to a plant. Indeed, as plants are often subject to damage, a brain would be a liability.

> **UNDERGROUND NETWORK**
>
> **In the movie *Avatar* (2009), all the plants on the alien planet are in communication with one another through the roots of trees. This is not far from the actual situation on Earth, with an underground network of fungi allowing plants of many types, from trees to grass, to communicate through the transfer of chemicals.**

Plants have a modular design, with roots, leaves, flowers and branches repeated in the same pattern. Consequently, a plant can lose up to 90 per cent of its body and still regenerate – there are no unique vital organs to be lost. A brain that could be bitten off by an ant or an antelope would reduce the plant's chances of survival.

Although plants don't have nerve cells, they do produce neurotransmitters such as dopamine and serotonin, chemicals

PLANT TELEPATHY

In 1966, CIA polygraph expert Cleve Backster connected the polygraph to a plant in his office – for no particularly good reason. He found, he claimed, that if he imagined setting fire to the plant, it produced a surge of electrical activity that registered on the polygraph. It seemed as if plants not only felt fear, but could be mind-readers.

Backster went on to hook up polygraph machines to other plants including lettuces and even picked fruit and vegetables, such as onions, oranges and bananas. He reported that plants responded to the thoughts of people nearby and, in the case of people already familiar to them, over a distance. Putting his CIA training to good use, Backster even claimed to have got a plant to pick a criminal out of a line-up. The plant was present at the murder of another plant (it had been stamped on) and picked out the killer from a line-up, producing a surge of electricity when the culprit appeared. They also responded to violence towards other types of organism, producing a stress response when live shrimp were dropped into boiling water or an egg was cracked in front of them.

The story of Backster's telepathic plants featured in a bestselling book, *The Secret Life of Plants*, by Peter Tompkins and Christopher Bird, published in 1973. Some of the impact endures, with people talking to their plants or playing them classical music and claiming it improves the plants' wellbeing.

Other scientists have not been able to replicate Backster's results with the polygraph. Most plant scientists were outraged at the damage they felt the book's wild claims did to the reputation of their work, and some complain that research into plant responses – provocatively called 'plant neurobiology' – is not taken seriously.

that in animal brains are used to send signals. No one knows the function or operation of neurotransmitters in plants. Plants can also respond to anaesthetics, a discovery first made in the 19th century. Exposing them to ether (an early anaesthetic) prevents them from photosynthesizing, stops seeds germinating, and deters mimosa leaves from curling when touched. The way that anaesthetics work on plants is still not fully understood.

CHAPTER 21

Do we all see the same colours?

We agree that grass is green and the sky is blue – but do we all see colours in the same way?

It's a tricky question, as we can never know what it is like to be someone else. It's an issue that strays into the territory of philosophy as well as psychology, but the mechanics of seeing depend on straightforward physics.

Ingredients of seeing

When you look at an object, light reflected from or emitted by the object enters your eye through the pupil and falls onto photoreceptive cells on the inside back surface (the retina). There are several types of cell - rods and cones are the ones responsible for detecting light. Cones are responsible for colour vision; rods help us see at night by distinguishing between light and dark in high resolution.

Cone cells are sensitive to a range of wavelengths, but respond most strongly to light of a particular wavelength; they transmit signals - electrical pulses - to the brain. The closer the light is to their peak wavelength, the more pulses they produce and the stronger the signal they send. The brain puts together all the information from the rods and cones to build up a colour picture of the world. It works in much the same way as a television or computer screen, which builds up a complex image from a vast collection of individual pixels. The brain combines information from the six million cone cells in each eye, giving us high-resolution, three-dimensional vision.

How things look

When light (of any colour) falls on an object, there are three things that can happen: the light can be absorbed, reflected or transmitted. The light is absorbed if the wavelength of the light coincides with the frequency of vibration of electrons in atoms in the substance. The impact of the light is to heat the substance slightly, so light energy is transformed into heat. A black object absorbs all light energy - which is why if you leave

SEEING CELLS

The human eye has about 120 million rod cells and 6 million cone cells. Rod cells are very sensitive to low light levels and can be triggered by a single photon (particle of light energy). They enable us to see in dim light, but don't detect colour. Nocturnal animals, such as owls, have an even higher ratio of rods to cones, giving them very good night vision.

We can see shapes because we have lots of cone cells packed together closely in a small area at the back of the retina. Cones can detect colour because there are three types that specialize in different wavelengths of light. Called S-cones, M-cones and L-cones, they are most sensitive to, respectively, short, medium and long wavelengths of light. For people with normal vision, this maps to blue (S), green (M) and red (L) light.

Everything looks grey in the night and in shadow because there is not enough light for the cones to work. The light reflected by objects is exactly the same – it still has a range of colours – but we can't detect the colours because the light intensity is too low for the cones to work.

a black object in sunlight it heats up more than a similar-sized white object (because white objects reflect all light).

If light falls on a substance that doesn't have electrons with a matching frequency, it excites the electrons briefly and then is re-emitted as light. If the substance is opaque, the energy radiates out from the surface – it's reflected.

If the substance is translucent, the energy is passed to adjacent atoms, exciting those electrons, and so on, passing from atom to atom until it comes out the other side of the object – so the light is transmitted. A little radiates back the way it came, so green glass both appears green when you look at it (reflected light) and allows green light to shine through it (transmitted light).

A BIT ABOUT COLOURS

Light is part of the electromagnetic spectrum: it's energy that travels in waves. The electromagnetic spectrum includes energy with widely different wavelengths (the gap between the peaks or troughs of waves). The spectrum ranges from radio waves to gamma waves. Radio waves have the longest wavelength; there can be 100 km (62 miles) between waves. Gamma rays have the smallest wavelengths, about the size of the nucleus of an atom. The part of the spectrum we can see is called visible light. Light of different colours has different wavelengths – red light has a longer wavelength than blue light. In terms of physics, the only difference between visible light and radio waves from a distant star is the wavelength. Our bodies respond to light, radiowaves and gamma rays in very different ways, however.

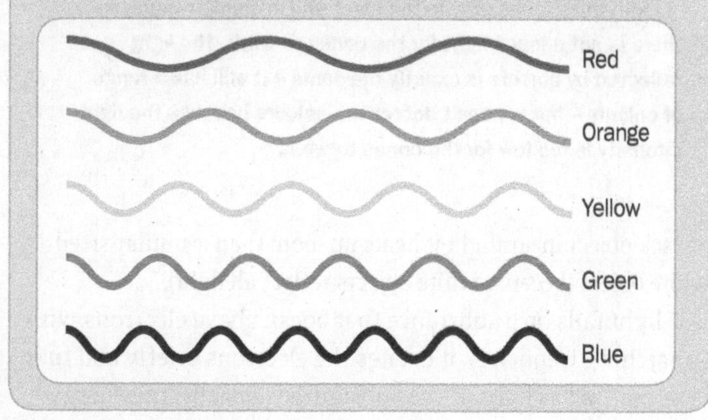

Colours are real

It is certainly the case that you will see light of the same wavelength as the same colour. Every time you see an object that is reflecting yellow light, with a wavelength of 570 nm, you will see the same colour. The physical properties of the object and the light remain the same, no matter who looks at them, or whether there is no observer at all. So if we define colour not as something perceived but as the wavelength of light involved, we do all see the same colours. But we might not all experience the same colour in the same way.

The physical construction of the cone cells and the nerves that connect them to the brain, and of the brain itself, is not the same in all people. The nerve transmissions are carried out in the form of chemical reactions – these are always the same and cannot be altered. But the photoreceptors might be slightly differently tuned – perhaps your L-cones (good for seeing red) are most sensitive to light with a wavelength of 564 nm, but mine might be most sensitive to light with a wavelength of 567 nm. This will make me more sensitive to orangey reds. Not only that, but the colours one person 'sees' in their mind's eye might appear different from the colour someone else 'sees'.

More colours and fewer colours

Although we all have the same way of hearing sounds, some people can hear sounds of a higher or lower frequency than others, so the abilities of our ears and other sensory organs are not identical.

Some known vision problems mean that certain people – those with various types of colour blindness – don't see some of the colours others see. Someone might see, for example, greens as shades of brown. If they have learned to label something 'green', it may take a long time for them to become aware they don't see green in the same way as other people – but eventually it usually becomes obvious that other people can see different shades and call things green or brown when the person with colour blindness can see little or no difference between them.

Another variant is that a few people are 'tetrachromats' – they have four types of cone instead of three. The extra type of cone is most sensitive to colours between red and green (in the yellow/orange range). Tetrachromats also see colours more clearly or vibrantly than other people. Some animals have four types of cone as well and can see a broader spectrum of wavelengths than we can. Some insects, such as bees, are able to see in the ultraviolet. Some snakes, such as rattlesnakes, can see in the infrared.

What colour is the dress?

In 2015, a photo of a striped dress went viral on the web. Some people saw it as white and gold and others, looking at exactly the same photo, saw it as black and blue. Even experts could not account confidently for the difference in the way people saw the dress. Dr Jay Neitz, who researches colour vision at the University of Washington, suggested that perhaps the photo had been taken in bluish light and some people unconsciously compensate for that and see the dress as white, while others don't compensate and see it as blue. But it was a tentative suggestion, not a solid explanation. What the photo did make clear was that we don't necessarily all see the same thing when we look at the same object.

Again, this is not the same as agreeing that something is green, but having a different colour experience when looking at it. The way our brains make sense of the world is chemically identical yet we all know that people respond differently to many things – from liking and disliking flavours to finding something painful or pleasurable. There are a few flavours and smells that some people can detect and other people are completely unaware of. It seems that our brains are capable of presenting a different experience from the same stimulus – but it's very difficult to tell whether we could be seeing different colours. Though we might all agree that blood and tomatoes are both red, do you see them as what I would call blue? If you looked at a picture of a blue tomato, you would realize it is not the red that you associate with tomatoes. What you can't tell is whether it is the colour I see when I look at normal tomatoes.

WORDS MATTER

It appears that the number of colours people distinguish in a rainbow depends partly on their eyes and partly on the language they speak and the number of colour-words it has. Isaac Newton, who first explained that the spectrum is produced by splitting white light, could only distinguish five colours, but settled eventually on seven to describe it. This was because he was attracted to an Ancient Greek theory that maintains there is a connection between colours, the seven planets known in the 17th century, musical scales and the days of the week. People who live in cultures with fewer colour-words tend to distinguish fewer colours clearly.

All the same to us – or not?

In 2009, Neitz and his colleagues carried out an experiment on male squirrel monkeys, which have only two types of cone cells so can see fewer colours than we can. The male monkeys cannot normally see red and green – they can't distinguish them from a neutral background – although female monkeys can. The scientists inserted a human cone cell gene into a random selection of the male monkeys' cone cells. The gene was transported by a virus which infected some of the cone cells, adding the gene and turning them to cells that could detect red. After five months, the monkeys with the infected cone cells were able to distinguish red. This means that even

'I think we can say for certain that people don't see the same colors.'

Joseph Carroll, Medical College of Wisconsin

though their brains were not equipped at birth to distinguish red, they could adapt to do so. Neitz claims it means that there is no predetermined pattern to how brains build up perceptions of the external world, at least as far as colour vision is concerned.

CHAPTER 22

Will we ever find a cure for cancer?

In the UK, there are around 450 deaths from cancer each day. Will we ever manage to eradicate this disease?

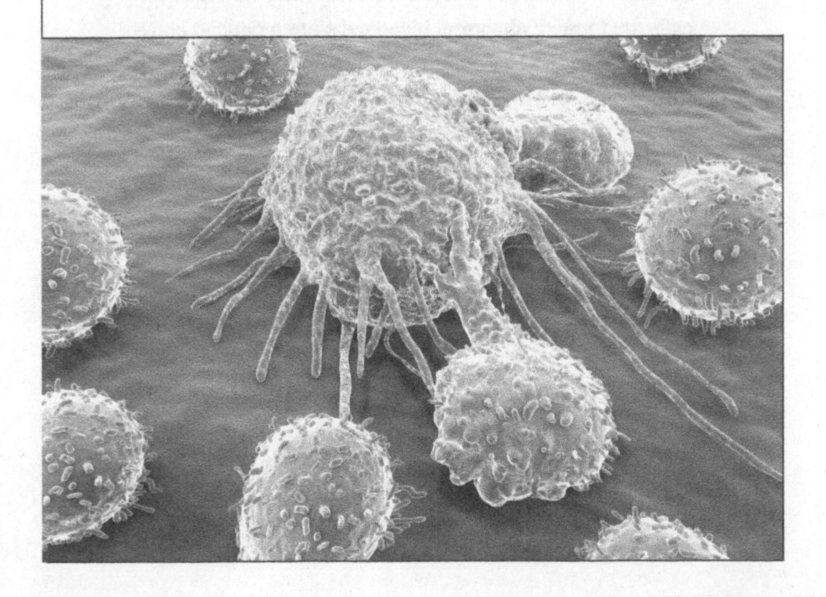

Out of control

The human body is made up of cells organized into tissues and organs. The cells are of different types, each suited to their different functions. An adult has around 100 trillion cells (100,000,000,000,000). Cells normally divide in a controlled way: when we need new cells in order to grow larger (when we are children), to repair damage, or to replace old cells that have worn out, cells copy their genetic material and divide in two.

CATEGORIES OF CANCER

There are around 200 types of cancer that affect different parts of the body. Cancers are grouped into five categories depending on the type of body tissue they start in.

- **Carcinoma** is cancer that starts in the skin or in the linings that cover internal organs.
- **Sarcoma** begins in tissues that hold the body together or support it, such as bones, cartilage, muscle, fat and blood vessels.
- **Leukemia** begins in the bone marrow where new blood cells are formed; abnormal blood cells are produced and released into the bloodstream.
- **Lymphoma and myeloma** begin in the structures of the immune system, such as the lymph glands.
- **Brain and spinal cord cancer** start in the tissues of the nervous system.

Sometimes dividing cells make a mistake. The genetic material, DNA, is not properly copied. This is called a mutation. Mutated cells often die, but sometimes they do not. They can then start multiplying – creating more defective cells. Cancer occurs when cell duplication runs out of control. The defective cells continue to divide when new cells are not needed. These can develop into a lump or growth, called a tumour. Cancers of the blood (leukemia) don't form tumours, but the extra cells build up in the blood vessels or bone marrow (where blood cells are created) and cause problems. The tumour can grow undetected for some time – even years – depending on where it is in the body and how quickly it is growing. Some tumours are benign (non-cancerous), but they can still be life-threatening if they put pressure on other structures, such as those in the brain.

A tumour only becomes malignant (cancerous) when it develops the ability to spread to other parts of the body, usually when tumour cells break off and are carried in the blood or lymph system. They can then develop into secondary tumours elsewhere in the body.

Causes of cancer

There are many factors that can trigger cells to reproduce unnecessarily. It can be caused by external influences on the cell, such as tobacco smoke or radiation. It can come about as the body ages – our cells are more likely to make mistakes as we age, and errors in cells accumulate.

Some people inherit genes which increase the risk of a particular cancer. It can happen entirely randomly, because cells make mistakes when duplicating themselves. And it can result from a combination of several of these factors.

It takes many mutations to lead to cancer. The body's quality-control mechanism usually detects defective cells and causes them to self-destruct. But if the mutation involves this checking system and mistakes get through undetected, or if the mutation prevents the cell from self-destructing, the faulty cell can successfully reproduce and cancer may develop.

Part of you

The difficulty for medics treating cancer is that the tumour is formed from the body's own cells. With infectious diseases, there is a pathogen – a disease-causing organism – which is alien to the body. The body's natural defence mechanism, the immune system, exists to attack such alien organisms and destroy them. It's our own personal army of special cells to fight disease. We can help the immune system with medicines such as antibiotics (see *Is this the end for antibiotics?* page 135), or even supply ready-made antibodies to fight a specific disease. But cancer presents a tricky problem. Although the immune system will generally destroy diseased or 'broken' cells, cancer cells are recognized by the immune system as part of the 'self' and are not attacked. (Indeed, auto-immune diseases, in which the immune system destroys the body's own cells, cause serious illness.)

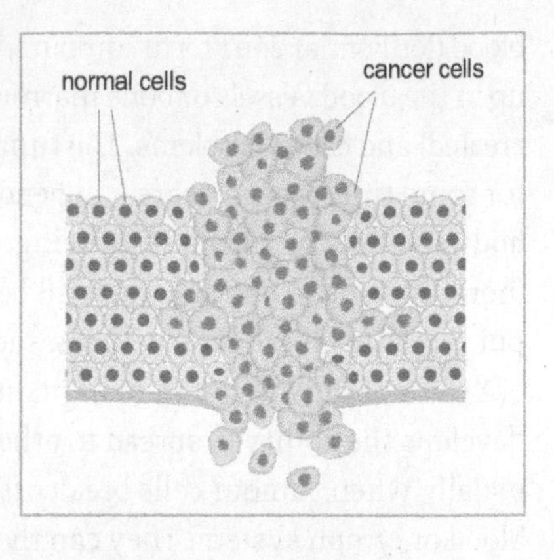

Treatments

Treatments for cancer are often unpleasant and distressing. In the past, they were even worse. Until the 20th century, the only treatment was surgery – to cut out the tumour. This sometimes worked,although it was an agonizing experience in the days before anaesthetics.

Today, the tumour might still be removed surgically (with the patient anaesthetized) and then the patient treated with chemotherapy or radiotherapy. But often cancer can be treated without surgery.

Chemotherapy, in particular, can cause very unpleasant side-effects. Chemotherapy targets cells as they divide. Usually it is given as chemicals that travel through the whole body. It can be given as an injection, a drip or as capsules to be swallowed. As cancer cells reproduce more rapidly than most other cells, they are the ones that are damaged most by chemotherapy. However, other cells which reproduce frequently are also affected, including skin cells, hair follicles and cells in the lining of the gut. This is why people having chemotherapy often lose their hair and suffer an upset stomach – the cells in the scalp and gut are collateral damage.

Radiotherapy works by directing a large dose of radiation specifically at the tumour cells. The blast of radiation breaks the long threads of DNA molecules that make up the chromosomes. This makes it impossible for the cells to reproduce and they die. Radiotherapy is focused specifically on the tumour so that it doesn't damage cells elsewhere in the body. Some radiotherapy is given by mouth or by injection into a vein, where it travels around the whole body until it reaches the tumour, where it accumulates.

> ***'Yet – when the dreadful steel was plunged into the breast – cutting through veins – arteries – flesh – nerves – I needed no injunctions not to restrain my cries. I began a scream that lasted unintermittingly during the whole time of the incision – and I almost marvel that it rings not in my Ears still! so excruciating was the agony.'***
>
> Fanny Burney, writing about her mastectomy to Esther Burney, 1812

Both these methods work by disrupting the DNA in cancer cells. Cancer cells lack the mechanisms normal cells have to repair damaged DNA, and this is the feature that traditional cancer treatments exploit. When normal cells are damaged by radiation, they halt the cell cycle until the damage has been repaired. Cancer cells carry on regardless: they continue to reproduce, but the catastrophically damaged cells produced can't function properly and rapidly die. The details of the process were unknown when radiotherapy and chemotherapy were first developed – it was clear that the treaments worked, but not clear how.

Cancer cells can continue to mutate, however, and in the process may develop resistance to the drugs used to treat them, and even to other drugs that they have not yet been exposed to. For example, cancer cells may make adaptive changes that enable them to pump drugs back out of the cells.

Better methods

Medical research is constantly looking for improved ways to treat cancer. There are several promising developments, though it's too early to say whether there is a miracle cure around the corner.

'The evidence emerging from clinical trials suggests that we are at the beginning of a whole new era for cancer treatments. Some of the most common types of cancer seem to be treatable with immunotherapy. Overall, cancers of the lung, kidney, bladder, head and neck, and melanoma cause about 50,000 deaths a year, or around one third of cancer deaths.'

Peter Johnson, Professor of Medical Oncology, Cancer Research UK

One approach is to target the cancer's ability to hide from the body's immune system. New immunotherapy drugs aim to do just that, re-educating the immune system to enable it to 'see' and destroy cancer cells. Early results announced in 2015 included a British study in which 58 per cent of patients with advanced skin cancer had their tumours considerably reduced using this method, and in 10 per cent of cases the tumours were destroyed.

Another possibility is to turn off the gene which encourages tumour cells to replicate. Scientists are working with naturally occurring strands of genetic material called messenger RNA, which can selectively block a single gene. This process is known as RNA interference (or RNAi). Research in 2012 found that when a particular gene was blocked in a patient with a type of leukemia, the cancerous cells stopped replicating and returned to normal, non-dividing, white blood cells. It's possible that a form of RNAi molecule can be found to interfere with other cancer cells. The trick lies in finding the protein that is causing the problem in each case and then blocking its production with an appropriate RNAi.

Exploiting a lack

Cancer cells often lack a gene called p53. This lack can possibly be exploited in treatment. The p53 gene is part of a cell's defence mechanism against viruses, but some viruses produce proteins to inactivate p53 in cells. The virus then enters the cell and multiplies. The way viruses work is to hijack the cell and instruct it to make copies of themselves. When the cell is full of copies of the virus, it bursts and the viruses tumble out to invade neighbouring cells.

One virus, an adenovirus, has been engineered so that it can only live inside cells that *already* lack p53. This means it will only be able to target cancer cells and won't invade healthy cells. If the virus is put into a tumour, it will do the job of killing the cancer cells but will leave normal body cells intact as it is unable to enter them. This approach is undergoing clinical trials in the USA and is already used in China to treat lung cancer.

Blocking chemicals

Because cancer cells divide and spread so rapidly, they need a good blood supply to nourish the cells and remove waste. Cancers generally co-opt local cells, prompting them to develop into blood vessels when the tumour is about 1 mm (0.04 in) across. It's possible to spot chemical markers in cells that are about to undergo this change, and also to identify and block the action of chemicals that prompt the development of blood vessels. Blocking the production of these chemicals or their action might be a way to starve tumours into submission.

By studying the DNA of cancer cells, scientists are beginning to reveal similarities between cancers of distinct types. This already helps medics to describe and identify specific cancers, and to give a prognosis for patients; it also paves the way for new treatments. If medical investigation into cancer cells can reveal their specific weaknesses, it should be possible, in the future, to find treatments to which the cancer is especially vulnerable by exploiting those weaknesses. It would lead to an approach in which each patient's cancer is profiled at the molecular level and attacked with carefully targeted treatments.

Around half of the population living in the economically developed world will develop cancer at some point in their lives and one in four will die of it. A cure for cancer would save a great deal of suffering. Less distressing treatments would be one of the most welcome medical developments.

CHAPTER 23

Could intelligent machines take over?

Intelligent machines taking over the world is a familiar science fiction scenario. But is it a real threat?

In order for machines to take over, they would need to be capable of independent thought and reasoning, beyond the relatively straightforward programming of most of today's computers. In a word, they would need to be intelligent.

Intelligent machines – known as 'artificial intelligence' (AI) – do not yet exist in a form we would recognize as rivalling human intelligence. So the danger is not just around the corner, but it might not be far off. As some technologists and philosophers have pointed out, the time to think about the problem is before it can happen, just in case it needs to be avoided at an early stage of development. It's too late when the machines are already lined up against us, flexing their steely muscles.

Step 1: Make the machines

Since the first robots appeared in the middle of the 20th century, people have worked towards AI.

Before the challenge of making AI can be addressed, we need to decide what counts as intelligence. There's no universal agreement on what it is, but phrases such as independent thought, creativity and theory of mind (the beliefs, desires and intentions that are used to understand why a person acts in a certain way) are bandied about. Neither is it clear where the boundaries would lie: Could an AI deliberately deceive? Could it make guesses? Could it have feelings? Would it be conscious?

The imitation game

The first test of whether a machine could be called intelligent was devised long before there was an AI to test. The computer pioneer Alan Turing described a test he called the 'imitation game', and which we now call the Turing test. It involved allowing a human to have a typed conversation with a hidden computer. If the person could reliably tell they were talking to a machine, the machine failed the test. But if the person was fooled into believing they were talking to another human, the machine passed the test.

We haven't yet developed anything that passes the Turing test. But as we all deal with machines that speak to us, from call-centre phone systems to the talking 'digital assistants' provided with some smart phones, we are getting more used to the idea. We are perhaps also more savvy than our predecessors – we might be harder to fool now.

There is a lot of relatively low-level AI around already. The actions of the computer-controlled players in a sophisticated video game are decided by an AI program, for instance. Digital personal assistants such as Apple's Siri and Google Now use AI to decode and answer the questions we ask using natural language (normal speech, rather than computer code). But they are a very long way off the true intelligence that enables humans to interact with one another and understand the hidden complexities of language.

A sentence such as 'time flies like an arrow; fruit flies like a banana' is difficult for a computer to parse – there is no clue that 'flies' is to be treated as a verb in the first case and a noun in the second. Language is littered with such traps: if 'buffalo milk' is milk *from* buffaloes, then, following the same logic, 'baby milk' is milk *from* babies. The complexity of the neural networks in our brains enables us to bring years of experience and a wealth of context to bear on such phrases. As far as we know, no computers come close to this level of complexity yet.

It is a misconception, though, to assume that any artificial intelligence must mimic human intelligence. Mimicry is not something we demand of other machines. All our mass transport devices use wheels, a solution to movement that is not found anywhere in nature. It's possible – even likely – that AI could develop along lines which do not parallel human intelligence.

AI in the cloud

Key to developing AI is giving systems a way of learning. To be truly autonomous and intelligent, they will need to be able to step beyond the knowledge provided with their programming to absorb new knowledge, learning from each interaction, from new circumstances and consequences of actions. If this learning is constrained to the experience of any single AI, it proceeds slowly. But if AI systems pool their experience and learning in the cloud, several or many systems can share and draw on a much wider bank of knowledge.

AI friends, helpers and lovers

Science fiction movies have postulated AIs we could fall in love with, have sex with, make friends with or that would be house servants. Some organizations are already working on AI personal care assistants that help elderly people with routine tasks, answering a need for care assistance in countries with an ageing population and too few young people to provide sufficient care.

Japan faces one of the most challenging demographic outlooks, so it is no surprise that it leads the world in developing care robots. It's a market that is likely to expand in many parts of the world. Robots will probably begin by taking on practical tasks such as delivering a medicine trolley or making sure fire exits are always clear and emergency equipment is in place and working. This will free human care staff for more interactive tasks. But robots are increasingly used for more personal care, such as turning patients or helping them to wash their hair – a robot with 24 fingers developed in Japan can already wash hair at least as effectively as a human helper.

Popular in Japan, too, is a furry seal-shaped robotic companion that simulates the responses of a living pet, showing appreciation when petted. Many owners enjoy this

AI companionship, but some critics feel it strays too far into more controversial territory. It is such a new development that there is no research yet into the long-term psychological effects of people growing emotionally attached to AIs.

Step 2: Rely on the machines

It is difficult to believe that anyone would willingly set out to develop an AI that would try to take over the world, enslave or annihilate humans, or embark on any of the other destructive paths featured in the typical science-fiction Armageddon scenario. These major mishaps would most likely be unintended consequences. We would no doubt start with benign aims, such as making life easier and more pleasant for ourselves, using resources more efficiently, protecting the climate, caring for people who need a lot of assistance, making information readily available, diagnosing and treating illnesses more accurately and reliably, and so on. But danger lies in success.

Bill Joy, co-founder of Sun Microsystems, has suggested that the more useful we make machines, the more we will come to rely on them. After a while, it will be impossible to turn them off. We are already over reliant on computerized systems. Turning off the computers that run the banks or air traffic control, for example, is unthinkable even now.

We have seen unfortunate consequences from the speed with which computerized decision-making takes place. In 2010, the so-called 'flash crash' saw the Dow Jones Industrial Average drop by about 1,000 points. Supercomputers trading at lightning speed, with no human overseers, caused the value of stocks to fall because they sold too quickly. It is easy to see that as AI becomes better and better at making decisions, surpassing human decision-makers, we are likely to allow it to make more and more decisions. After all, why would we

– or any commercial organization – choose a seemingly less reliable decision-making system (people) over a better one (AI)? It looks almost inevitable that AI will take over an increasing share of our tasks once it becomes widely available.

Step 3: Oops!

Computers, though, are only as good as their programming. Even if they are able to learn completely independently, they will not (as far as we know) develop a conscience, compassion, an ethical system, emotions or consciousness unless they have been programmed to do so. It's easy to see how a system without these human safeguards could get out of hand. Bill Joy cites a scenario in which an AI created to optimize the production of paperclips could decide to commandeer all available resources to achieve its goal, even, potentially, taking atoms from human bodies to make into paperclips.

Death by paperclip is not a particularly serious threat, but many other unintended and undesirable consequences have been posited. One suggestion is that AIs might choose, entirely

CONSCIOUSNESS AS EMERGENT PROPERTY

There is a theory that consciousness emerges naturally from neural networks – it doesn't have to be created but comes about naturally, much as weather is an emergent property of atmospheric air, water and geology. This doesn't explain what consciousness is, but explains how it might come about. It also explains the 'hive mind' or group consciousness of animals that act together, such as ants and bees. If the theory is correct, consciousness might arise automatically in any sufficiently complex computer system. It might even be there already. It need not be a form of consciousness we would immediately recognize.

logically, to exterminate humanity on the grounds that people are detrimental to the planet. It might be possible to prevent that happening – adopting Asimov's laws of robotics would be one way (see box below).

There are other, less obvious dangers. The Industrial Revolution mechanized many time-consuming and dull tasks previously performed by people, with the result that workers in some fields lost their jobs. We have seen the same with computerization and the advent of robots in manufacturing. AI could remove even more jobs, including many that currently involve expert knowledge, such as work in medicine, the law, education, and professions such as architecture and

ASIMOV'S LAWS OF ROBOTICS

Science fiction author Isaac Asimov first set out his three laws of robotics in the short story 'Runaround' (1942). These are:

1. **A robot may not injure a human being or, through inaction, allow a human being to come to harm.**
2. **A robot must obey the orders given to it by human beings, except where such orders would conflict with the First Law.**
3. **A robot must protect its own existence as long as such protection does not conflict with the First or Second Laws.**

Asimov later added a fourth rule, which came before the others:

0. **A robot may not harm humanity, or, by inaction, allow humanity to come to harm.**

It has often been suggested that these would be quite good laws to govern the production of any AI in the real world.

scientific research. AIs taking on more of our work could be either devastating or liberating, depending on how society handles the change. Large groups of people could be left out of work, unmotivated, and perhaps prey to mental illness. They might be placated by drugs and entertainment, encouraged to take up hobbies and kept satisfied to prevent social unrest. In some dystopian visions, the mass of people who are no longer needed to work might even be wiped out, or forbidden to have children, by an elite with no need for them.

Hans Moravec, who founded the robotics programme at Carnegie Mellon University in the USA, believes that AIs will eventually succeed human beings. He envisages us keeping them relatively well controlled for quite a while, but it's a battle he feels we must lose in the end. Perhaps it will be cyborgs – human/machine combinations – who take over, but he considers the writing will eventually be on the wall for human beings once we let the AI genie out of the bottle.

CHAPTER 24

What's the difference between a person and a lettuce?

It's hard to believe it, but we share genetic material with salad.

You probably don't look much like a lettuce, so you may wonder exactly which of your genes are the ones you share with that salad vegetable.

A bit about DNA

The information that forms the blueprint for any organism is coded into its DNA (deoxyribonucleic acid) – the extremely long molecules found in the cells. Each strand of DNA is called a chromosome and is made up of sections called genes. The entire set of genes that makes up the 'recipe' for an organism is called its genome.

The genome comprises a complete instruction set for building and operating the organism's body. The instructions are copied into virtually every cell of the body, and tell each cell what to be and what to do – be a bone cell and harden with minerals, or be a nerve and carry nerve impulses, and so on.

Gene genie?

Each gene instructs the body to make a particular protein. Proteins are responsible for all the activities that take place within the body, from digestion to growing and fighting disease. Each cell only follows the instructions to make the proteins it requires – it only reads the appropriate bits of the DNA. The other bits of DNA are coiled away where they can't be reached and are switched off.

We tend to look at chromosomes rather in the way we look at underground stations marked on a map: we only take notice of the stations, but there is essential track in between. Genes whose function we recognize only make up about 2 per cent of the total length of the chromosomes. Scientists are still exploring what the other 98 per cent does. It's likely that one function performed by 'the track' – the non-coding DNA – is to tell each cell which of the proteins it has to make. As every cell has the full instruction set, this is important information. You wouldn't want cells in your eyes producing digestive enzymes, for example.

What those genes do

Many cells in different organisms have to carry out similar or identical functions. Although you might not look much like a lettuce, your cells carry out many processes which are the same as those a plant needs to do. Certain cell processes are the same in all or most organisms, including the way cells divide and the way they take energy from glucose in the process of cell respiration.

ALL IN CODE

The structure of DNA is a double helix with links between the two strands, rather like the rungs of a ladder. Each 'rung' consists of a pair of nitrogenous 'bases' (bases are alkaline substances which form salts when mixed with acids). These are called 'base pairs' and are always adenine-thymine and guanine-cytosine – they can't combine differently.

The base pairs occur in groups of three, called a 'codon'. As each pair can start with any of the four, and there are three pairs in a codon, there are 4 × 4 × 4 = 64 combinations in total. A sequence of codons gives the recipe for a protein – the information the gene needs to provide in order for the cell to make the proteins it needs. Proteins are made up of amino acids. There are easily enough codons for each of the 20 amino acids used in constructing proteins to have its own code. There are also specific 'start' and 'end' codons that mark the beginning and end of a gene.

So although most of your body is nothing like a lettuce, it does have similar functions and processes to other plants and animals, particularly other mammals. This explains the larger portion of our genome that we share with other animals.

SIZE DOESN'T MATTER

The human genome has 46 chromosomes, with around 3.2 billion base pairs. By contrast, yeast has only 12 million base pairs, some poor bacteria have just over 100,000 and the fire ant has 480 million. But don't feel too superior about your huge genome: the marbled lungfish has 130 billion base pairs.

Animal	Proportion of genome shared
Chimpanzee	90%
Mouse	88%
Dog	84%
Zebra fish	73%
Chicken	65%
Fruit fly	47%
Roundworm	38%

The reason we share so much genetic material with other organisms is that we have evolutionary ancestors in common with them. In general, we share more genes with organisms from which we have more recently diverged during evolution. Humans and chimpanzees had a common ancestor as recently as around six million years ago, so the genetic differences that have emerged since then are those which make chimps and humans different animals. We have to go further back in time to find common ancestors with other mammals, fish, birds, insects and plants. As a consequence, more genetic changes have accumulated. Our last common ancestor with sharks might have lived 290 million years ago – there has been a lot of time for humans and sharks to evolve differently, losing and gaining different genes along the way.

There's not, though, a direct link between evolutionary distance and genetic difference. The differences come about as both lines of development go through genetic changes through mutation. So the differences between a mouse and a human are the sum of the differences between the human and the common ancestor and differences between the mouse and the common ancestor. If one line is developing more slowly, there might be fewer accumulated genetic differences than we would expect. Further, some genes might change in the same ways if the organisms are subject to the same kind of environmental pressures. Either way, in some regards you are and will remain similar to a lettuce, a mouse, a shark and everything else that has evolved.

CHAPTER 25

How will the universe end?

The end of the universe will come – eventually.

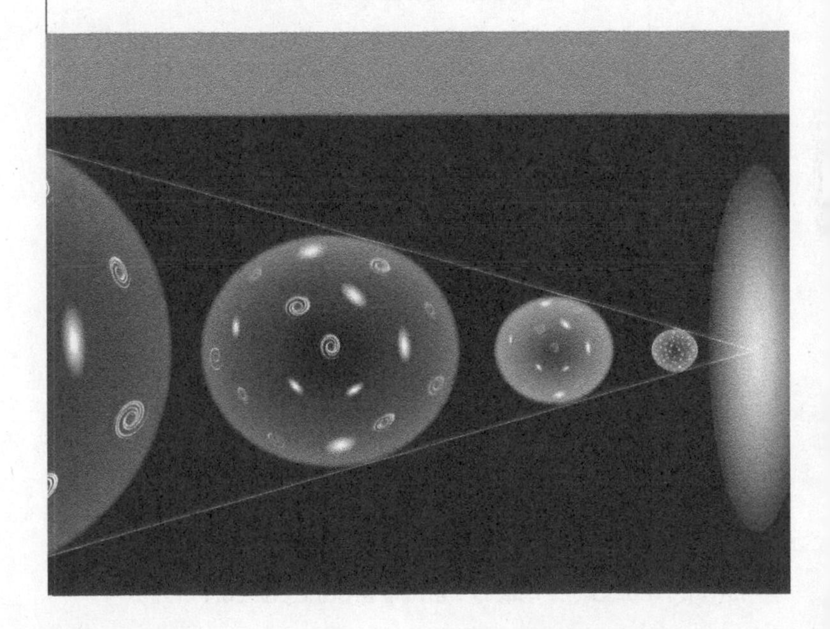

Scientists have a pretty good idea of how our universe started, which they have derived by working backwards from what we can observe of the present universe using the laws of physics as we currently understand them. But they are not agreed on how it will end. It could go out with a bang or a whimper.

We know that the universe is expanding. But no one knows how long it will go on expanding, and what will happen in the end. Traditionally, there have been three possibilities:

- it keeps expanding, going faster and faster in an 'open' universe
- it keeps expanding, but at the same or a slower rate in a 'flat' universe
- it stops expanding, goes into reverse and collapses in on itself in a 'closed' universe.

None is very nice, but luckily none will happen for about 20 billion years or so: it's not top of the list of things you need to worry about.

Those are the familiar scenarios which have been around for a while. But another possibility has been added to the list, and it's one that could happen this afternoon, or next year, or in 18 billion years' time. In this scenario, the unstable universe will simply stop existing. You could worry about this, as it might happen in – or, strictly speaking, at the end of – your lifetime. But there's nothing to be done about it and we probably wouldn't notice it happening. It would all be over before you knew it.

We don't know quite enough physics yet to say which of these will happen, or if maybe there is something else we haven't thought of that is waiting around the corner.

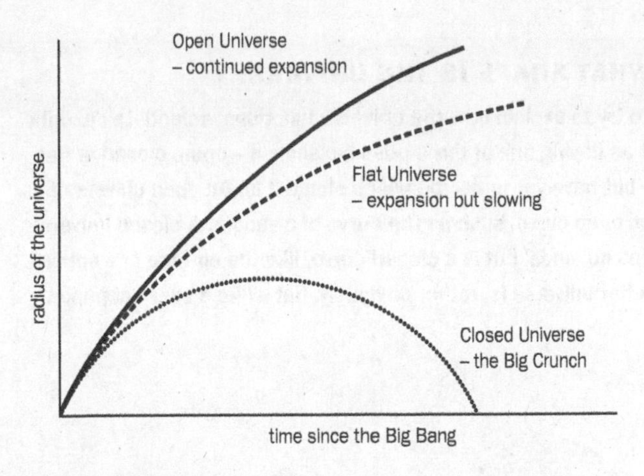

From the beginning

Current cosmological theory explains the beginning of the universe as a 'Big Bang', more properly called a 'singularity'. At this moment, all time, space and matter expanded into existence from an infinitely dense and small point. As no one can say where this point came from – and as 'where' has no meaning in this context – this still leaves the door open for supernatural explanations, such as a creating deity.

At this point, around 13.8 billion years ago, everything expanded massively, growing from the size of virtually nothing to huge in a millionth of a second.

In the first 10^{-32} of a second (that's 0.000000000000000 00000000000000001 of a second) it grew to the size of a grapefruit. At that point, its expansion slowed – but it was still pretty rapid: by the end of the first second the universe was around the size of our solar system.

In that first second, the first forms of matter and anti-matter appeared and mostly annihilated one another, leaving a surplus of matter in the form of quarks, electrons, photons, neutrinos and some other particles. The single force that was driving

WHAT SHAPE IS THE UNIVERSE?

To try to explain how the universe functions, scientists describe it as having one of three possible shapes – open, closed or flat – but have yet to decide which shape it is. An open universe has an open curve, such as the curve of a saddle. A closed universe has no 'ends' but is a closed curve, like the surface of a sphere. A flat universe is, rather obviously, flat – like a sheet of paper.

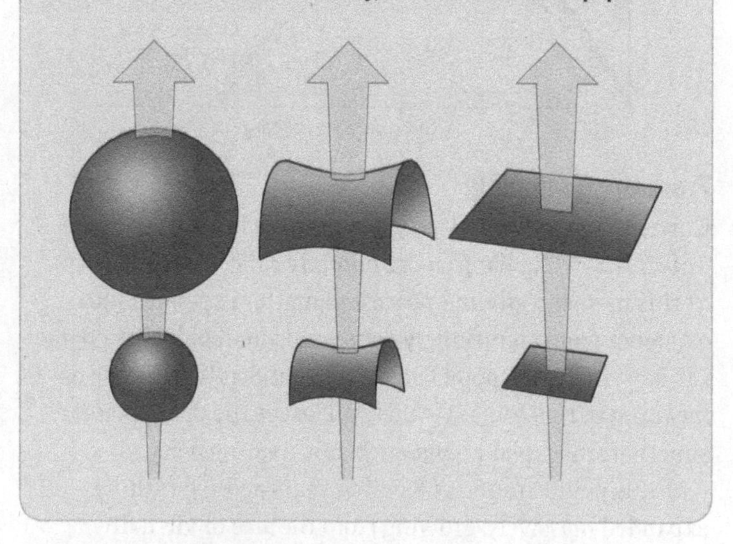

the universe split up to create some of the forces we now have, such as gravity, electromagnetism and the weak nuclear force. The particles started to smash together and form protons and neutrons. Although there were photons, the universe was still so dense that no light could shine – if there had been anyone there to see it, they would have seen nothing. A trip back to witness the Big Bang would be disappointing in so many ways.

How to make a universe

Over the next few seconds, as the temperature fell to only a billion degrees or so, protons and neutrons combined to form hydrogen and helium nuclei. After around 20 minutes, the universe was too cold and not dense enough for the formation of nuclei to continue – it was all over already. A hot soupy mix of particles continued for around 240,000 years.

By the end of that time, it was cool enough (about 3,000 °C) for atomic nuclei to start capturing electrons, so atoms started to form – matter, as we know it. By 300,000 years after the Big Bang, the universe was a fog of hydrogen (75 per cent), helium (25 per cent) and tiny traces of lithium. With particles clinging together in atoms, it was possible for light to shine, but nothing was there to emit light. The photons were just whizzing around, but there was nothing to see. We can still observe some of them now, as cosmic background radiation.

Then – not much happened, for about 150 million years. Finally, gravitational collapse caused the first quasars to form, and from 300 million years stars and galaxies entered the scene. This all happened as slight irregularities in the density of the spreading matter caused clumps and gaps to intensify. Gravity made the clumpy bits more intense, drawing the matter in them together. As matter gets closer, gravity works harder to draw it together and finally it collapses inon itself. As this happened, the collapsed matter became so hot and dense that hydrogen nuclei began to fuse together to form helium, releasing huge amounts of energy in a process known as nuclear fusion: the first stars were born. At last there was something to see.

These first stars were short-lived and super-massive (about 100 times the mass of the Sun) that soon exploded as supernovae. The debris reformed as new stars. Vast areas of matter drew together into galaxies, and galaxies into galaxy clusters, finally making the universe we have now. Our own Sun formed around 4.6 billion years ago, made up from matter recycled through many generations of previous stars over 8 billion years.

Where we are now

Currently, the universe is still expanding and cooling. It's a lot less dense, with about 10^{-26} kg per cubic metre, or 10^{-20} of a milligram of matter per cubic metre: that works out at about 16 hydrogen atoms per cubic metre of space. There's a lot of emptiness. The distribution and movement of matter through the universe is controlled by gravity pulling it together and some force – probably dark matter or dark energy – pulling it apart. This giant, cosmic push-me-pull-you sets the rate of expansion (or contraction).

Still growing

The Belgian priest and astronomer Georges Lemaître (1894–1966) was the first scientist to suggest the idea of an expanding universe, in 1927, working from Einstein's equations, but he had published in French and most astronomers missed it. His work was translated into English in 1931, and at that point he also suggested that if the universe is expanding it must once have been much smaller, what he called the 'Primeval Atom', with all matter in a highly compressed state. This expansion was confirmed by the astronomer Edwin Hubble in 1929 and subsequently became known as the Big Bang theory.

For a long time, it was assumed that the rate of expansion would gradually slow. Then, in 1998, data from the Hubble Space Telescope (named in honour of the astronomer) revealed that the universe is not only still expanding, but actually increasing its rate of expansion. This really threw a spanner in the cosmic works.

Georges Lemaître

The Big Rip

Many astronomers currently favour the Big Rip, a theory published in 2003 that would work in an open universe. In this scenario, the expansion of the universe will go on and on. While gravity pulls matter together, dark energy pushes it apart, increasing the space between matter. The speed of the expansion will become ever faster as the universe spreads further and further apart and gravity is able to put up less of a fight. Stars and planets will be torn apart. Finally, even atoms will be ripped apart. At some point in a finite period of time, the distances between things in the universe will become infinite. (As the universe started infinitely small,

GETTING SMALLER AS IT GETS BIGGER . . .

In a strange twist, as the universe keeps expanding, the part that is the observable universe – the bit we can see or receive radiation from – becomes smaller. It does not shrink in terms of its physical extent in kilometres, but the matter inside it reduces. Expansion pushes things currently on the edge of the observable universe further away from us, over the boundary into unobservable space. As the universe becomes bigger, we will see less of it.

dense and hot, and will end infinitely diffuse, that suggests that its finite state, its finitude, is limited. This is rather neat, and more pleasing than the idea that it will stay finite for an infinite period.)

There is a scary-looking equation for working out the time before the Big Rip happens:

$$t_{rip} - t_0 \approx \frac{2}{3|1 + w|H_0\sqrt{1 - \Omega_m}}$$

The result could come out at as little as 22 billion years, or as much as 35–50 billion years.

The countdown to disaster will start about 60 million years before the end, when gravity becomes too weak to hold galaxies together. The Milky Way will drift apart, our solar system (if it still exists) and others set free to wander at will – but not for the whole 60 million years. When there is only around three months to go, solar systems will dismantle themselves as their gravity will unravel. In the last few minutes, individual stars and planets will be torn apart, and in the last instant even atoms will be destroyed.

If you could live to see it, that means there might be a second or so when you get to float without gravity and before you personally disintegrate.

The Big Freeze

A flat universe would be one in which expansion continues, neither accelerating nor slowing, but plodding on indefinitely until matter is so dissipated and the universe so cold that everything grinds to a halt. The temperature in the universe will tend towards absolute zero (about -273 °C). Eventually, full entropy will be reached, which means that everything has fallen apart and matter is equally, thinly distributed over the whole universe.

What will it be like? First of all, the supply of gas needed for stars to form will run out. This will happen in around 1–100 trillion years, quite a long time in the future, with even the lowest estimate more than seventy times as long as the universe has lasted so far. As existing stars run out of fuel for their nuclear fusion reactions, they will die and the universe will slowly darken. Black holes will proliferate, but even they will deteriorate over time as they emit Hawking radiation, slowly eroding themselves to nothing. In around 10^{100} years, the universe will be a thin soup of virtually stationary electron-like particles.

ABSOLUTE ZERO

The term 'absolute zero' means zero Kelvin, which is equivalent to about -273.15 °C. It is the point at which matter contains no heat energy at all. Nothing moves, and strange quantum effects emerge. The coldest temperature that researchers have achieved is 0.45nK, or about half a billionth of a Kelvin.

Growing, growing – grown?

The possibility of a closed universe would give an entirely different outcome. In this case, the universe will reach a limit, and when it gets there it will collapse back in on itself. The reversal would be slow at first, but would gather speed as it progressed. At first the contraction would be fairly regular, but it would become increasingly uneven as matter became more concentrated into specific areas. Stars would explode and vaporize, and eventually even atoms would fall apart, rewinding the sequence in which matter appeared after the Big Bang. According to some theorists, the final stages of the collapse would be chaotic, causing massive distortions of space-time. Some even suggest that space-time would shatter into 'droplets', rendering all our ideas about time, distance and direction meaningless. The process of collapse could be complete in about 100 billion years from now.

If the universe collapses in such a Big Crunch, it might then spawn another Big Bang. One theory, called the Big Bounce, suggests the universe we are in currently is just one in a series of Big Bangs and Big Crunches. These could go on – and might have been going on – forever, with a series of finite universes bounded by an infinite series of expansions and collapses.

This is currently the least compelling scenario. Since the evidence for increasing speed of expansion emerged in 1998, there seems little reason to suppose that the universe will slow down and go into reverse.

The Big Slurp

The final scenario is the only one worth worrying about in the short term, but even then there is nothing we can do either to prevent or predict it. It rests on the suspicion that the universe is inherently unstable. This takes the mass of the Higgs boson and calculates that everything which is able to form in the universe now, from stars to life-forms, can do so only because the universe is teetering on the brink of stability. A tiny bubble of true vacuum could form at any point in the universe and expand exponentially in an instant – well, at the speed of light – wiping out everything in our current universe. We wouldn't see it coming or notice it happening.

> ***'Without warning, a bubble of true vacuum could nucleate somewhere in the universe and move outwards at the speed of light, and before we realized what swept by us our protons would decay away.'***
>
> Michael Turner and Frank Wilczek, *Nature*, 1982

Picture Credits

Corbis: 39 (Ikon Images)
ESA/Hubble & NASA: 34
Shutterstock: 7, 8, 13, 18, 23, 24, 25, 31, 32, 33, 45, 46, 51, 52, 58, 61, 67, 69, 71, 72, 76 top, 78, 81, 84, 85, 91, 101, 107, 111, 119, 121, 123, 126, 132, 133, 135, 141, 145, 148, 149, 150, 153, 163, 165, 166, 167, 168, 171, 183, 191, 194, 199, 207, 213, 218
Diagrams by David Woodroffe: 36, 49 (x2), 53, 55, 63, 76 bottom, 129, 158, 186, 214, 215